The Last Voyage of
the Whaling Bark *Progress*

ALSO BY DANIEL GIFFORD

*American Holiday Postcards, 1905–1915:
Imagery and Context* (McFarland, 2013)

The Last Voyage of the Whaling Bark *Progress*

New Bedford, Chicago and the Twilight of an Industry

DANIEL GIFFORD

McFarland & Company, Inc., Publishers
Jefferson, North Carolina

LIBRARY OF CONGRESS CATALOGUING-IN-PUBLICATION DATA

Names: Gifford, Daniel, 1970- author.
Title: The last voyage of the whaling bark Progress : New Bedford, Chicago and the twilight of an industry / Daniel Gifford.
Other titles: New Bedford, Chicago and the twilight of an industry
Description: Jefferson, North Carolina : McFarland & Company, Inc., Publishers, 2020. | Includes bibliographical references and index.
Identifiers: LCCN 2020018369 |
ISBN 9781476682150 (paperback : acid free paper) ∞
ISBN 9781476640075 (ebook)
Subjects: LCSH: Progress (Bark)—History. | Whaling—Massachusetts—New England—History—19th Century. | Whaling ships—Massachusetts—New Bedford—History—19th Century. | World's Columbian Exposition (1893 : Chicago, Ill.) | Exhibitions—Illinois—Chicago—History—19th century.
Classification: LCC VM465 .G54 2020 | DDC 639.2/80284—dc23
LC record available at https://lccn.loc.gov/2020018369

BRITISH LIBRARY CATALOGUING DATA ARE AVAILABLE

ISBN (print) 978-1-4766-8215-0
ISBN (ebook) 978-1-4766-4007-5

© 2020 Daniel Gifford. All rights reserved

No part of this book may be reproduced or transmitted in any form or by any means, electronic or mechanical, including photocopying or recording, or by any information storage and retrieval system, without permission in writing from the publisher.

Front cover illustration: "The Arctic Whaler *Progress*," by G.A. Coffin, from *"There She Blows" or the Story of the Progress*, 1893 (Arctic Whaling Exhibit Company)

Printed in the United States of America

*McFarland & Company, Inc., Publishers
Box 611, Jefferson, North Carolina 28640
www.mcfarlandpub.com*

To the descendants of
Captain Daniel W. Gifford

Table of Contents

Acknowledgments — ix
Preface — 1

One. New Bedford—June 8, 1892 — 5
Two. New Bedford, 1851 — 13
Three. Somewhere in the South Atlantic, 1858 — 23
Four. New Bedford, 1865 — 40
Five. The Arctic, 1871 — 51
Six. New Bedford, 1880 — 67
Seven. Chicago, 1892 — 82
Eight. Lake Ontario, 1892 — 96
Nine. Chicago, 1892 — 115
Ten. New Bedford, 1904 — 154

Chapter Notes — 171
Bibliography — 187
Index — 193

Acknowledgments

This book ultimately centers on questions of how we define and celebrate the work of a particular community. So perhaps it is no accident that writing it has led me to a community of amazing people, far more diverse and wide-ranging than I could have imagined at the outset. As always when I practice the historian's craft, the dedication of librarians and archivists who are at the center of our historical depositories never ceases to inspire me. Those that made this work possible include Nina Wright at the Westerly Public Library & Wilcox Park; Jodi Goodman at the New Bedford Free Public Library; Cheryl Ziegler at the Union League Club of Chicago; Don McPhee at the Peabody Essex Museum, as well as Jennifer Hornsby at the museum's Phillips Library; Louisa Alger Watrous and Maureen Smith at the Mystic Seaport Museum; David Hill at the American Numismatic Society; Armand Esai at the Field Museum, along with Nina Cummings, the Field's photo archivist; and Mark Procknik at the New Bedford Whaling Museum. Mark and his staff especially deserve acknowledgment for their insightful assistance during my time as a Scholar in Residence at the Museum, an honor for which I am enormously grateful. That residency proved pivotal to this project. The staff of the Harold Washington Library Center and the Racine Public Library were also instrumental in turning up the sort of forgotten layers of detail that made researching this story so satisfying. Closer to home, Mary Marlatt at the University of Louisville likewise answered my call for help. Thank you!

As early drafts began to coalesce, I was heart-warmed and awed by the willingness of the historical community to assist in the editorial and revision process. Jan da Silva at the New Bedford Whaling National Historical Park especially deserves accolades for spearheading a collection of volunteers who lent their insights and technical expertise to the project. These generous individuals from the Park included Mark Mello, Judy Roderiques, and Andrew Schnetzer. Akeia A.F. Benard, Curator of Social History at the New Bedford Whaling Museum, also graciously gave her time and insights to an early draft of this book. And I was especially overwhelmed by the detailed and hugely helpful input of Michael R. Harrison, the Obed Macy Director of Research

Acknowledgments

and Collections at the Nantucket Historical Association. As must be clear by now, the whaling history community has been enormously supportive of this endeavor, and I cannot thank them enough for their guidance and tutelage.

Working with the piles of records and edits generated by the process has likewise been an exercise of community-building and collaborative support. Special thanks must be given to Lara Kelland, Joshua Boydstun, Abby Glogower, Jodi Lewis, Fedja Buric, Jim Gibson, and my "History Dudes" Zac Distel and Monty Fields. Family across the generations of Giffords have also demonstrated their dedication to telling this story and have offered both encouragement and artifacts to see this through. Finally, there is the individual who crosses all these categories including researcher (or more accurately—researcher extraordinaire), editor, proofreader, sounding board, and tireless advocate. My husband Steve made this book possible in every sense. He deserves, and will always have, my highest levels of humble gratitude and love.

Preface

"Nothing in Sight. So ends."—Logbook of the whaling bark *Henry Taber*, November 13, 1868

"Heated the works. Employed boyling and other ship's duty. Nothing in sight. So ends."—Logbook of the whaling bark *Rousseau*, November 3, 1854

"Dropped our anchor at Clark's Point Lite and furled our sails. And so ends the Corinthian's voyage."—Logbook of whaling bark *Corinthian*, April 21, 1866[1]

The phrase "so ends" runs throughout whaling logbooks, but it also seems to resonate through time and memory. Whenever I encountered the phrase in the course of writing this book, I was reminded of many endings—for whaling, for whalers, and for an entire way of life, including for certain Giffords who lived in nineteenth century New Bedford, Massachusetts.

My family had always been aware of our whaling heritage, particularly the colorfully nicknamed ancestor "Bloody Dan," a New Bedford whaling captain. Captain Daniel W. Gifford was my great-great grandfather, and I could find the remnants of our whaling legacy throughout my grandparents' house—a barometer, ship model, scrimshaw, books—objects that in time found their way into my parents' home, and then eventually mine as well. The use of whaling artifacts was put to particularly good effect during grade school "show and tells."

Long before I ever took my first graduate class in American history, I rediscovered this ancestor through an exercise in family history, which involved retracing the voyages of Captain Gifford. It was in the process of this research in 1998 that I made a surprise discovery: He had taken a whaling ship to the Chicago World's Fair of 1893. The goal of moving the ship to Chicago was to create a whaling museum at the famed exposition's South Pond. It was a detail I found fascinating but did not particularly focus on, given the limits of my still-amateur research abilities at the time. What I did glean from the project, however, was a keen sense of metaphor. Captain Gifford was born

Preface

in 1839 and died in 1899—his 60 years tracking with almost perfect precision the rise and fall of New Bedford's whaling industry.

When I began my graduate studies in 2004, I returned several times to the World's Columbian Exposition (its official name, although most people just called it the Columbian Exposition or the Chicago World's Fair, as I will do here) and this time began to tease out more substance from the story that a whaling ship had traveled from New Bedford to Chicago to go on display. As I became more focused on museum studies in my doctoral program, I came to realize that the ship my ancestor had taken to the fair and converted to a museum—the whaling bark *Progress*—was also a larger metaphor for the whaling industry, albeit a particularly ironic one given that "progress" in 1890s New Bedford meant anything but whale oil and whalebone. Suddenly I was much more aware of questions about memorialization, historic memory, and myth-making within a museum context, and began to see the complexity that the *Progress*'s story contained.

Still, it took me some time to fully commit to telling this story. In a refrain familiar to any academic, life got in the way—a pair of museum jobs at the Smithsonian Institution, teaching, publishing my first book, the lectures and interviews that followed, more teaching—but always with New Bedford's fair-bound *Progress* niggling in the back of my mind. It took the fresh start of moving to a new hometown to shake me out of my revere and revisit the *Progress*.

Once my focus was reaffixed to the circumstances surrounding a whaling ship's strange freshwater journey to a world's fair, I couldn't help but see the sources and stories with a sort of double vision. Not only was I exploring evidence of the past, but I was exhuming a process that continues to generate debate and confusion to this day, something I explore more fully in the first chapter. So many times communities encounter the possibilities and pitfalls of trying to turn a dying industry in a museum piece. Often begun with the best of intentions, the process is fraught for even the best museum professionals. For the visitor, the end results can fall anywhere on the spectrum from uplifting to frustrating. The *Progress* is a fascinating case study in this regard—filled with surprising twists, interesting characters, and layers to which I have hopefully given justice.

Often the purpose of a microhistory like this one is to refract the larger stories of the past through a single historical narrative. In addition to the topics of memorialization and commemoration, New Bedford's suggestion of displaying a paeon to American whaling at the Chicago World's Fair had much longer and deeper roots than might be obvious on the surface. One nagging question of this project: Just how far back does one need to rewind

Preface

whaling's long chronology within American history? With any good superhero, there is an origin story. The elevation of an entire industry or trade to a heroic level in the American consciousness suggests the same need to trace the beginnings of the mythology.

Thus, this book is roughly divided into two halves, with the emergence of the *Progress*'s fair-bound journey only beginning after a necessary exploration of the years, decades, and shifting currents of history leading up to that moment. I found that origin story not in the Gilded Age New Bedford of the 1890s, but the antebellum New Bedford of the 1850s. Far from the New Bedford wharves from which the *Progress* began her journey were older stories that took place in the balmy waters of the South Atlantic and the frozen wastelands of the Arctic. And without stretching the comic book analogy too far, forays into earlier decades also help to illuminate if this tale would produce a villain or villains, and what their own origins might have been.

The result is a longer journey through history than simply 1893, the year of the Columbian Exposition. This is perhaps fitting for an industry that came to be defined by voyages that stretched on for years at a time. The first six chapters fully set the stage for the *Progress*'s world's fair story, which is told from beginning to end in the second half of the book over the remaining four chapters. Each chapter is anchored by a historical figure whose narrative helps explore the *Progress*'s voyage to Chicago, along with the whaling industry's rise, fall, and eventual conversion into a museum piece. The histories that propel these pages include the whaling industry's semi-religious sheen; marauding Confederate privateers; a dramatic Arctic rescue of 1,200 whalers, women, and children; and 200 nearly-drowned schoolkids in the Chicago River. In the final chapter, we can find legacies of the *Progress* throughout New England and Chicago.

What began as a tantalizing piece of family history has blossomed into a historical account of a whole community and way of life, rich in layers of hubris, hope, mythology, despair, and redemption. Yet in reconstructing this account, I never strayed far from that original sense of metaphor that hung over the entire enterprise of creating a whaling museum at the Columbian Exposition. This belief in the ability of objects to communicate stories is why we build and perpetuate museums. At their core, museums are created so that humans can gather, preserve, and interpret the very things that help us share and disseminate the stories we wish to tell.

But this process has so many opportunities for missteps, miscommunications, and miserable results. What objects? Whose stories? Who is doing the telling and who is doing the listening? What can seem obvious and inherent to one person can become distorted and irrelevant once received

Preface

and processed by someone else. "Exhibitions are fundamentally theatrical," writes Barbara Kirschenblatt-Gimblett, one of the museum studies field's top scholars, "for they are how museums perform the knowledge they create."[2] An 1890s museum dedicated to whaling and set on the marshy shores of the Columbian Exposition turns out to be no different.

One of the great satisfactions of researching and writing this book was rediscovering the surroundings, careers, and the complex choices of a small group of individuals who thought they were making a lasting impact on the world only to discover otherwise. History famously rewards the winners, and stories of failure often disappear into the historical dustbins, requiring our diligence and patience to unearth them. The arc of this story is ultimately a sad one, but it is an important one nonetheless.

So ends.

One

New Bedford—June 8, 1892

It seemed as if the entire harbor had spontaneously erupted into a palette of red, white, and blue. The colors gleamed in the bright sunlight and a stiff southeastern breeze tugged at the streamers and bunting that were generously affixed to nearly any object or edifice the decorators could find. The summer gusts also played with countless hats, jackets, dresses, and ribbons among the tightly-packed crowds who had congregated along the wharves by mid-morning. A few stalwart boys who sat at water's edge minded their fishing poles rather than the estimated 3,000 onlookers packed behind them. But they too could not keep their eyes off the gleaming hulk of red and white that sat just off to their right. The whaling bark *Progress* had also been festooned with bunting and streamers, which snapped in the wind from mast heads and yardarms. A large signal flag waving from the mainmast declared in bright, tall letters the name of the bark, although she needed no introduction to those who had gathered. The crack of American flags echoed off the façade of the New Bedford Cordage Company, before which the *Progress* lay in wait. Her hull had been painted a bright brick red. Her davits holding the six-man whaleboats gleamed white, as did the deckhouse. James E. Reed, the prominent African American photographer who had co-founded Headley & Reed on Purchase Street, ensured that the proceedings were photographed to his exacting specification and with his famous eye for detail.[1]

As the *Progress* was towed away from Rotch's Wharf, the crowd erupted into a deafening cheer, which gave way to song. After a turn of "Life on the Ocean Wave," Hill's brass band led the New Bedford throngs though the melancholy words of "The Girl I Left Behind Me." Harbor traffic—also bedecked in the decorative finery of signal flags and bright colors—added to the cacophony. Bells and whistles accompanied the multitude's rendition of the well-known tune: "Ye gods above oh hear my prayer, to my beauteous fair to find me; and send me safely back again, to the girl I left behind me." Ladies pulled out handkerchiefs to wave as the bark was pulled towards Buzzard's Bay.

By that blustery day in June 1892, New Bedford had been launching

The Last Voyage of the Whaling Bark *Progress*

One. New Bedford—June 8, 1892

whaling ships to every part of the globe for a century. Thousands of voyages had begun from the same wharves that the *Progress* now passed on her way down to the Bay—the Central Wharf, the Taber Wharf, the Atlantic Wharf. Tens of thousands of men had unmoored from New Bedford docks, knowing the passage of time that marked distance from loved ones would typically be measured in years, not weeks or months. An incalculable number of people in New Bedford had watched whalers disappear onto the horizon, wondering if they would ever return. And yet, no single departure of a whaling ship had ever generated the sort of holiday atmosphere that surrounded the departure of *this* particular whaling ship. This departure was different. It was special. And it carried with it the legacies, dreams, and memories of an entire community.

The *Progress* was not going to sea to hunt for whales, as so many whaleships had done before. She was going to Chicago to be displayed like a crown jewel for all the world to see. She was going to the Columbian Exposition of 1893, and New Bedforders were sure she would be one of the most popular attractions at a world's fair that promised to be the event of the decade, perhaps even the century. She would tell American visitors from Maine to California a story about New Bedford's whaling heritage and history, and she would remind the world how New Bedford had once lit the globe and lubricated the industrial revolution with whale oil. The cheers and huzzahs and whistles gave voice to a civic pride that brought many New Bedforders to tears that day. The *Progress* would mark their place in history.

Chicago—April 14, 1900

Spring was slowly working its way into Chicago's frigid bones. The day had been mild and pleasant, and that night a full moon rose and shone down from clear skies. The skybound sentry cast a pale light over the lapping waters of the southwest shore of Lake Michigan. It was quiet save for the calls and beating wings of various ducks, gulls, and herons. They would occasionally alight on the now-unrecognizable husk of the *Progress*.

She rested on her side in the mud of the Harbor of the Calumet, a

Opposite: "Last Farewell to New Bedford." The *Progress* on her day of departure from New Bedford, with a portion of the nearly 3,000 spectators on hand for the occasion. Although uncredited, the photograph was likely taken by the prominent African American photographer James E. Reed, who had been contracted to record the festive day. From *"There She Blows"* (Chicago: Arctic Whaling Exhibit, Co. 1893).

The Last Voyage of the Whaling Bark *Progress*

stench-filled point where the mouth of the Calumet River vainly tried to discharge an accumulation of industrial waste and sewage into Lake Michigan. But the current of the river was not strong enough, and inky pools of oil, livestock blood, and muck collected and shimmered in the moonlight. While steel mills, chemical plants, and packinghouses upriver belched and buzzed as busy hives of modern labor, the atmosphere here was decidedly more graveyard-esque. Surrounding the *Progress* were the moribund bodies of other vessels in varying states of decay and disrepair. Here was the derelict schooner *Mary E. Dykes*, damaged in a storm and now poking out of the fetid waters. Over there, just a few feet away from the *Progress*, was the *John A. Dix*, a Civil War–era cutter that was falling apart bit by bit.

The motley collection had all long ago been picked over for anything of worth, but the *Progress* especially had proven a boon for scavengers over the years. The copper from her hull had disappeared early on. Most of the usable wood went next, taken by men and women who lived on the margins and

The remains of the *Progress* at the mouth of the Calumet River. Photograph by E.C. Greenman, circa 1902. Greenman's grandfather helped build the whaleship in Rhode Island during the 1840s. Launched in 1842, she was originally named the *Charles Phelps*. While Greenman was in Chicago he visited what was left of the whaleship (courtesy of Westerly Library & Wilcox Park).

One. New Bedford—June 8, 1892

needed the easily burnable kindling to stay warm. The remaining shell was rank and rotting, part of an inconvenient assemblage that made navigating the Calumet difficult for modern ships.

This was where the *Progress*'s story ended, on her side in the foul, frigid waters around Chicago. It would be nearly two years longer before fire and dynamite would finally finish the job, and the few scraps of remaining wood—long-since shed of their gleaming red paint—would settle into the muddy bottom. Only the demolition team and flocks of startled birds would bear witness to this ignominious finale.[2]

In between these two moments is a rich tapestry of expectations and disappointments, assumptions and failures that carry a deep resonance today. The sentiments of the whaling industry in the 1890s echo across modern communities of coal miners and steel workers, newspaper journalists and video store owners. Those who find themselves in a dying industry are often beset by more than just questions about economic security and hope for the future. They also begin to ask questions about their legacy and place in the larger narrative of American history. Will they be forgotten? Will they be understood? Will they be pitied or celebrated? Who gets to decide?

To the extent a community of workers can control anything in the midst of a fading industry's decline, it is control over their own sense of identity and importance that often bubbles up to the forefront of our political discourse and public debate today. There is more at play than simply wanting a job to last another paycheck or another year or another generation. There is a strong desire not to be dismissed or erased. Writes Ryan Burg of the phenomenon: "We speak of dying industries and organizations, but we say that jobs are 'lost,' as if they have been misplaced. It sounds strange to describe a job as 'dead....'"[3] No worker wants to see their industry's tombstone inscribed prematurely.

Thus, it can be an unwanted and heavy burden when it falls to a community to shepherd its own legacy and commemoration when a way of life and livelihood fades out of existence. And yet, this happens with more regularity than we might realize. Each time an old factory is turned into loft apartments or a riverfront is given a new lease on life as a tourist destination, questions hang in the air about how much to memorialize the work that came before. It is up to those in the community—often living alongside the very laborers (or their decedents) whose work is now relegated to the past—to decide the answers to those questions. Sometimes a plaque, sculpture, or series of black-and-white photos provides the link; other times the path to honoring those memories is more elaborate and thorough. Many museums tell the stories of industries, trades, and workers who Americans have come to see as either fading or gone. Such museums dot the American landscape, particularly in the hard-hit areas we now commonly refer to as the "Rust Belt." The topics of these museums range from coal and steel to carousel horses and parlor organs.

This more complex and multi-layered task of building a museum to an industry was the strategy the people of New Bedford settled on in the early 1890s. For them, the answer started with repurposing an authentic whaling bark. Not only would this be converted into a museum dedicated to whaling, it would proudly be displayed at the Chicago World's Fair. The narrative of the *Progress*'s journey as found in the two scenes described above is a perfect example of the roads taken, and not taken, in the quest to tell a story about a community of workers. Yet bound within the story of the whaleship's exhibition at the Columbian Exposition are other stories that began long before 1893, and stories that continue to this day. It is also the story of a dying trade and lost traditions. It is the story of a city in transition between two economies. It is the story of memorialization and commemoration; memory and the process of declaring a once-thriving industry dead, even as it still has workers and families who depend on it. It is certainly a story about civic pride, but perhaps too a story tinged with darker threads such as presumptiveness and betrayal.

The journey of the *Progress* also straddles two of the nineteenth century's most important cultural and economic milestones—whaling and world's fairs. It is easy today to forget how impactful both were on American lives and the American economy. Whaling was understood in its day to be a magnificent driver of both dollars and innovation. Moreover, whaling generated outsized impacts on the entire country, relative to its geographical footprint. Every part of America was touched by whaling, but only a few places in the still-young nation actually pursued the trade. The physical space of the whaling industry was confined to a few seaboard towns, Nantucket and then New Bedford foremost among them. But as an industry that helped lead a new nation into the industrial age, and placed America upon the world economic stage, whaling's influence cannot be denied. At its peak, it was the fifth largest industry in America, a spot held by manufacturing today—no small coincidence for an exploration of dying ways of life. And whaling did not quietly grow into an economic juggernaut. Instead, it exploded—expanding by a factor of fourteen between 1826 and 1850.[4] Nobody blinked an eye when Senator William Seward described whaling in 1852 as America's "favorite enterprise."[5]

If whaling helped the United States usher in an age of economic and in-

One. New Bedford—June 8, 1892

dustrial might, the Columbian Exposition of 1893 was the megaphone used to announce America's arrival as an international powerhouse to the rest of the world. Although the United States had hosted its first major world's fair in 1876 Philadelphia as part of a centenary celebration, the Chicago World's Fair was on an altogether different scale. Beginning from, and trying to outdo, the legacy of Paris' Exposition Universelle of 1889—which gave the world the Eiffel Tower—Chicago tripled the size of Paris' fairground. World's fairs had always been conceived and designed as carefully organized showcases of industrial production, technological progress, and modern innovation, going back to London's Crystal Palace in 1851. By 1893 the scope of fairs had expanded into dozens of massive, white palaces dedicated to all the accomplishments of humanity and Western civilization especially—manufacturing, mining, transportation, agriculture, fine arts, and so on. Chicago invited nations, states, and companies to show the world their wares. And while nearly 50 nations participated in the fair, nobody doubted that the Columbian Exposition was really all about the United States and her miracle city of Chicago. "Many argued that the exposition was a clarifying moment in contemporary history," writes one scholar of the fair, "waking up people on both sides of the Atlantic to the 'power, progress, and importance' of the United States."[6]

Where these two stories—whaling and fairs—came together, perhaps even collided, was the *Progress*, and even more important was the idea of creating a whaling museum at the 1893 world's fair. As noted, it is a fraught proposition to declare someone, or someone's heritage, to be obsolete and a thing of the past. It can be a complicated process to turn an occupation into a museum piece. When President Trump used his first speech to Congress to declare "dying industries will come roaring back to life," he was in part tapping into the deep-seated desire in the American psyche not to give up on an industry too soon. For every eulogy about the American auto industry there is a Super Bowl ad promising that Detroit will rise like a phoenix. For every lament about the brutalities of e-commerce or the global economy, there is a feel-good story about independent bookstores and record shops making a comeback. There is importance to getting a community's story told and even greater importance to getting it told right and at the right time. It is what makes the story of the *Progress* both so fascinating and so resonant today.

The history that is revealed when we exhume the *Progress* from her watery, muddy grave returns again and again to that belief in the power of an industry and community of workers to craft their own narrative. So many individuals thought that the *Progress*, and the objects she contained, would tell the story *they* envisioned. Instead, the *Progress* revealed stories of light and dark, pride and despair, in combinations few could have foretold. There are lessons contained within that can guide our very contemporary debates and discourses about the place of work and the expendability of labor in modern America. The *Progress* is speaking to us over a hundred years later about how to value a community and their livelihood. All we need do is listen.

Two

New Bedford, 1851

James T. Almy was going back to the drawing board ... literally, in this case. A few years earlier in March 1847, New Bedford had graduated in its incorporation from town to city. Though perhaps a seemingly insignificant bureaucratic distinction today, the upgrade to city status was a testament not only to New Bedford's population growth, but also its increased economic importance. It was therefore unsurprising that one of the city council's first ordinances was to establish a city seal, so that they could officially stamp their growth onto every future document pertaining to city business. The charge to create the new city seal had gone to James Almy, a 22-year-old New Bedford engraver. Almy's design was accepted and served New Bedford for the next several years. When it debuted, it was described in the local newspaper as "a view of the Lighthouse at the entrance of the harbor, and several whaling ships are seen in the distance."[1]

There was only one problem—the lighthouse that Almy had so prominently displayed at the center of the proud city's new seal was, in fact, all wrong. As the story would later be recounted: "This seal had a serious error. The artist had located the lighthouse at the south end of Palmers island instead of the north end. The seal passed with this error until 1851 when a new drawing was made."[2] Now four years later, Almy essentially had a "do over" when thinking about how to immortalize the city he called home. It remained a challenge for the now 26-year-old to design a seal that befitted this prosperous and forward-looking locale—one that had sadly been saddled with unflattering nicknames like "Oil City," which did little to highlight its beauty and sophistication.[3] Almy's task was once again to give visual form to the sense of importance and self-worth that coursed through the wharves and streets of one of America's up and coming great metropolises.

The engraver simply had to step outside his shop on Union Street to be inspired. Union Street was the east/west corridor that ran to the wharves. Water Street was a similar thoroughfare which ran north/south, and both streets served as the center of New Bedford's economic and retail activity, particularly the "Four Corners" where Union and Water intersected. In other words, Almy's engraving shop was located at the heart of New Bedford's transformation.[4] Articles about New Bedford's prosperity and wealth had been circulating for several years in popular magazines. "There is scarcely a town in our country, of equal importance, about which so little has been said, by the book-makers, as New Bedford," began one widely-circulated piece in 1845, which ended with the reminder, "Let it be remembered that the wealth which we have seen, as belonging to the people of New Bedford, has been accumulated by its inhabitants. It has not been the result of foreign labour, or the capital of non-residents invested in the business of the place.... [It] has but few parallels in this, or any other country." Other pieces in America's prodigious output of periodicals touted similar themes, and collectively they documented a range of options that Almy surely considered when thinking about what to highlight about New Bedford: an advanced public-school system; paved and well-kept city streets with lighting; progressive poor relief; a modern fire department. And the source of this wealth—whaling.[5]

Almy's ill-fated first design did little to reflect these hallmarks of New Bedford's rising star, beyond a nod to whaling. The foregrounded, if inaccurate lighthouse took up the central, visual position of the original seal and four whaleships were scattered throughout the harbor, although only one was particularly clear and detailed. The town itself was relegated to a tiny sliver on the right. Across the top was the Latin "Nova Bedfordia Condita A.D. 1787," or "New Bedford founded in 1787." And along the bottom curled "Civitatis Regimine Donata, A.D. 1847," to designate when the city achieved its much-coveted city status.

When Almy submitted his redesign in 1851, almost all these details had been changed, with only the founding descriptions in Latin across the top and bottom remaining intact. The alterations reveal much about how Almy and the city's leaders saw their home and their place in the world. One of the changes was to the skyline of the city itself. In 1847, Almy had shown this as a triangular spit of buildings poking into the water at a distance. The new design also showed the city skyline at a distance, but now it stretched across the horizon from end to end. This time Almy had added in contrasting details in order to communicate growth and prosperity. The left end of the skyline starts with a few scattered buildings, representing farmlands and rural outskirts. Then, as the eye moves from left to right, the city builds in density and height, climbing as it goes so that it becomes a thick conglomeration of no less than 30 structures and steeples stacked like blocks on a hill. Boston may have been Massachusetts' original "city upon a hill," but clearly Almy was willing to give the capital a run for its money on this count.

The center of the seal presented the most significant changes. The of-

Two. New Bedford, 1851

fending lighthouse has been converted from the most prominent feature of the seal to one that is scarcely recognizable—a mere stump of a structure in the middle of the harbor. The whaleship, instead, is now the largest element, double the height of the lighthouse and dwarfing everything else in the image. Although now the whaler has been joined by other vessels including a steamboat, there is no doubt that the whaleship is the most important part of the scene. It seems clear that this is where Almy wishes to draw your eye. A teeny pair of onlookers survey the busy (and some might argue cluttered) scene from the bit of land that fills the bottom curve of the seal.

All the changes were approved by the City Council, which could breathe a sigh of relief that their city seal was no longer an exercise in visual misrepresentation. The Council documented their newly reminted endeavor with a seal-specific ordinance in the city record. It pointedly noted that the new seal now correctly presented "a view of the Northerly extremity of Palmer's Island with its Light House...."[6]

One additional change was also made to the seal—one that was not mentioned in the official record, but nonetheless spoke volumes about New Bedford's direction and frame of mind by the 1850s. Almy's original seal included the motto "*Lucem Diffundens.*" With the redesign, it became "*Lucem Diffundo.*" The former, connected to the prominent image of a lighthouse, translated to "diffusing light"—a relatively straightforward explanation of the

The original 1847 city seal (left) and revised 1853 city seal (right), were both created by New Bedford engraver James Almy. Changes include the diminishment of the lighthouse, the increased visual placement of the whaleship, and the modified city motto. *City Documents, Reports on Committee of Finance, etc.* New Bedford, 1850 (left) and 1852 (right) (courtesy New Bedford Whaling Museum).

The Last Voyage of the Whaling Bark *Progress*

lighthouse's function. The lighthouse might stand as a metaphor for other things, but the overall connection between subject and verb is unambiguous.

However, the new Latin motto translates more broadly to "I diffuse light" or "We diffuse light." This makes the seal much more complicated. Who, exactly, is the first-person conjugation referring to? Someone or something has suddenly become much more active in this new verb tense, and thus Almy's new seal has to make it obvious who is taking charge of this new, more forceful action. Given the overall demotion of the lighthouse in the new seal, it seems clear that Almy has shifted his focus away from it, including its potential metaphorical interpretations. In his new seal, the I/we referenced ceases to be a lonely lighthouse sentinel and instead becomes a reference to the seal's two most prominent features in the redesign—the whaling ship and the town itself.

Who originated or decided on the motto remains a bit of a mystery, with some suggesting it was the mayor's idea.[7] In a cosmopolitan city like New Bedford it was possible that someone, even Almy himself, had taken a fancy to gas company tokens circulating in Europe from the 1840s which declared *Lucem Diffundo Per Orbem* or "We Light the World." The fanciful tokens also included a torch-bearing lion straddling the globe. Minted for the "Societe D'Eclairage Par le Gaz," a pan-European association of gas companies founded in 1844, the tokens were ubiquitous and could be found in cities as diverse as Koblenz, Rennes, Naples, and Abbeville. Did one or more of these tokens drift into the path of a traveling New Bedforder councilman who saw the potential of the phrase? Of course, Latin was also widely taught in antebellum America among the educated classes, so it could be the phrase was simply selected by one of New Bedford's well-schooled native sons or daughters.

Whatever the origin, the phrase *Lucem Diffundo* stuck and remains the city's motto to the present day. In its longevity it also ended up being something of a Rorschach test to the citizens of New Bedford. There were the obvious religious undertones. Matthew 5:14 pairs "You are the light of the world" with the equally famous "city upon a hill" metaphor that John Winthrop had deployed so skillfully during Massachusetts' founding. Some have suggested the phrase had connections to abolitionists and the Underground Railroad, while an early twentieth century town librarian used the phrase to promise enlightenment and knowledge through books and educational library programs to the citizens of New Bedford.[8]

But one of New Bedford's definitive early historians, Zephaniah Pease, looked to put alternative interpretations to rest in 1925, and declared that the motto is about one thing, and one thing only: "The motto of the city

Two. New Bedford, 1851

was 'Lucem Diffundo,'—'We diffuse light.' It was the light of candles and whale oil lamps. Petroleum had not been discovered, and the uses of electricity were undreamed of."[9] Indeed, most sources have agreed with Pease in recognizing the motto's connection to the city's unique output—oil and candles—both the result of a complex economy and industry based on hunting whales. New Bedford was the nation's largest producer of three major products that pushed back the darkness and the night: whale oil, sperm oil, and spermaceti candles.

Whale oil came from rendering the blubber of certain whales—right whales and bowhead especially—into a product that eventually came out of the process viscous and hued with a range of browns, ambers, and yellows, depending on the quality of the blubber. Seal oil was much less commonly produced but largely used the same methods, generating a product that was white and bright in spite of the similarities of the process. Whale oil was known throughout the world as a dirty fuel, but a fuel nonetheless. When burned it produced a weak, smoky, and fishy-smelling flame that the poor and urban lower classes often used as a light source. In central Massachusetts, whale oil lamps were abundant in the majority of nonfarming households, augmented by the addition of cheap "looking glasses" or mirrors that helped amplify the light. Refilling whale oil lamps and cleaning up the smoky residue they left each night was a common chore among mid-century housewives.[10] When in 1855 the *Boston Chronicle* alerted readers to both the size of New Bedford's coffers and the source of its success, the headline was decidedly succinct and unflattering: "Rich and Oily."[11]

Much more desirable, and thus much more expensive, were sperm oil and sperm (spermaceti) candles. It is a surprisingly forthright and frank term to describe the milky substance produced by the sperm whale. However, the location of spermaceti wasn't genital—or gender-based. Although sure to still generate a snicker from teenage boys, the terms "sperm oil" and "sperm candle" are based on a misnomer. Spermaceti is found in the whale's head, not the reproductive organs. And there is a lot of it—a single large sperm whale could yield 500 gallons of the substance when harvested by whalers; some went as high as 1,000 gallons or more. Spermaceti was literally harvested by the bucketload, with buckets lowered by pole or rope through a carved hole in the head. Herman Melville described the fruits of this repetitive labor as reminiscent of "a dairy maid's pail of new milk."[12]

What consumers bought was a final product that took this raw material and processed it into liquid form—sperm oil—or the solid form of spermaceti candles. Unlike the oils produced from rendering the blubber of other whales, sperm oil was largely clean-burning and odorless. Spermaceti candles were bright and smokeless, perfect for illuminating a parlor for an evening of reading, sewing, or needlework. So good was the light, it eventually became a convention of measurement. Writes Philip Hoare: "The very unit of light, the lumen, was measured from the pure white spermaceti candle, one candlepower being equivalent to the burning of one hundred and twenty grains of wax per hour."[13] Benjamin Franklin was a big fan of spermaceti candles, writing that the product could "afford a clear white Light, may be held in the Hand, even in hot Weather, without softening [sic]; that their Drops do not make Grease Spots like those from common Candles; that they last much longer, and need little or no Snuffing."[14] When a young Frederick Olmsted took an ill-fated trip through the American South in the 1850s, he complained bitterly about the lack of comforts he took for granted back home, including hot water in the morning, a clean bed, and a parlor "glowing softly with the light of sperm candles."[15]

As Olmsted's protestations suggest, such products were also largely confined to the homes and neighborhoods of the wealthy—the cost of illuminating with sperm oil was roughly double that of whale oil and illuminating with spermaceti candles was triple the cost of the cheaper alternative. Still, people swore by them: "Their transparent whiteness renders them particularly applicable for parlor use," declared the distinguished chemist Campbell Morfit in 1856, "and though of higher cost than the ordinary kinds of candles, they are, considering their superiority, scarcely more expensive."[16]

This ubiquity in American homes was certainly a boon for New Bedford, but even the most refined parlor could only burn through so much whale product each year. More market applications were required. Fortunately, there were lighthouses. The U.S. Treasury Department's Lighthouse Board ordered tens of thousands of gallons of sperm oil annually, because of its clear and bright light. Although the department regularly experimented with other, cheaper forms of illumination—including cotton seed oil and rapeseed oil—it was hard to beat the results delivered by New Bedford's prized product. And New Bedford made sure to remind buyers of that fact should they be tempted elsewhere, at one point calling these potential alternatives oil choices "luminous humbugs."[17]

Indeed, having tried other possibilities, a Treasury Department official concluded in 1851 that nothing should be given "a preference over sperm oil, which, in my opinion, is the best material that can be used for producing lights in light-houses." The average Atlantic lighthouse burned through 35 gallons of sperm oil per lamp each year. Great Lakes lighthouses burned slightly less annually because they were not kept alight when icepack ceased shipping traffic for the season. Multiply this by the average nine and half lamps per

Two. New Bedford, 1851

American lighthouse, and again by the nearly 300 lighthouses that dotted the American sea- and lake-coasts in the early 1850s, and a picture emerges of just a single consumer—the United States Government—burning through approximately 95,000 gallons of oil per year.[18] There is something intriguingly cyclical in imagining whaling ships heading to sea and returning home guided by the lights produced from the very sperm oil they were pursuing.

The uses of whaling's light-generating products spread elsewhere. Some of the first city streetlamps were fueled by whale oil, driving back the night along dangerous lanes and thoroughfares. In the first half of the 1800s it was common in major cities and large towns to see globes or four-sided boxes atop wooden lampposts, from which lit whale oil lamps would emit light. London invested heavily in this technology as early as the 1730s when the number of oil lamps—called parish lamps—increased to fifteen thousand. The prominent stench of so many burning whale oil lamps did little to improve London's reputation. New York got its first lamps under British rule in 1752 when three were installed around City Hall in a drive to conquer the darkness.[19] Or at least attempt to. The light from whale oil street lamps had one wag suggesting that better illumination could come from "five and twenty full grown lightning bugs" while another suggested that whale oil lamps "seemed as if they were mourning for the loss of the moon."[20]

Still, other cities saw the civic and commercial value in attempting to combat darkness through whale oil, or for posher neighborhoods, in using the more expensive sperm oil streetlights. Pittsburgh added whale oil streetlamps in 1816; New Orleans had theirs along Canal Street by the 1790s. Washington, D.C., had a few whale oil street lights by 1801, but little money to operate them. Congress erected their own lamps on the Capitol grounds but saved money by only lighting them when Congress was actually in session.[21] Even New Bedford herself was not immune to controversies arising from ill-tended and poorly-illuminative whale oil street lamps. "Much complaint has been made," wrote the mayor in his annual address of 1852, "and justly as I think, with regard to our Street Lamps ... we who boast ourselves as the diffusers of light to all the world, are in some danger of becoming ourselves benighted."[22]

New Bedford sat at the center of all of this, from parlor candlesticks to luminous lighthouses to lampposts dotted throughout the urban landscape. The extent to which New Bedforders—and the leaders of her whaling industry especially—saw themselves as bringers and bearers of light to the world cannot be overstated. Certainly, whaling produced a whole range of other products—baleen, ambergris, industrial lubricants, watch and chronometer oils, and even fish, seal, and walrus oils. But none of these were put forward as the lifeblood of New Bedford by the phrase *Lucem Diffundo*. Not only did the sentiment of spreading light feature prominently in Almy's city seal, but it infused the way the industry saw itself.

It is no accident that in *Moby-Dick* Melville inserts a "member roll" of mythological and religious figures who serve as brothers to his whaling crew. What he ultimately is writing is a laundry list of figures who serve as heroes of light and defeaters of darkness. Melville lists Perseus, who is literally conceived from a golden shower of light, and who then goes on to use a mirror and reflected light to defeat Medusa. St. George is almost always seen as the personification of light killing darkness, a white knight on a white horse vanquishing some variation of a black, dark, reptilian beast. Vishnu is a sun god, his chakra depicted as a disk of light used to banish the darkness. "I myself," says Ishmael, "belong, though but subordinately, to so *emblazoned* a fraternity [emphasis added]." The use of the word "emblazoned" signals that the unifying thread of this particular fraternity is their shared role as "bringers of light."[23]

It was a belief that was further shaped and enhanced by the fact that most of New Bedford's ruling elites were members of the Religious Society of Friends, or Quakers, a faith that ran deep with metaphors about light. One scholar of early Quaker sermons writes, "In the sermons, light/dark imagery is so pervasive that it is impossible to understand other clusters of metaphor without understanding the early Quaker vision of light and dark."[24] "Believe in the Light so that you may become children of the Light," was an oft-repeated phrase in the Quaker faith, so much so that one historian noted that "the phrase 'inner-light' has also become inseparably attached to them [Quakers] and their successors." Disputes between groups of Quakers often came down to "New Lights" versus "Old Lights." Darkness too played an important role in the Quaker religious lexicon, with an emphasis on moving out of the darkness or dispelling the darkness. Quaker founder George Fox often spoke of Christ in substantial terms, specifically about Christ's ability to lead followers out of the "shadows, and out of death and darkness, up into the light...."[25]

The very same products that dispelled the actual darkness—lamp oil, candles, lighthouse fuel—thus were infused with larger connotations and meanings for those raised in a faith built upon the battle between light and dark. George Fox commandeered the metaphor of candles well before New Bedford started producing them, suggesting that "man's Natural Spirit [i.e., inner or inward light] *is* the candle of the Lord, but it is Christ who lights that candle ... [emphasis in original]."[26] Other ministers encouraged each congregation to "shake herself from the dust and put on her beautiful Garment, that her Light may shine forth as a Lamp that burneth, that thro' the Brightens

Two. New Bedford, 1851

No. 53 Pine, near William St,
NEW YORK CITY..................NEW YORK.
OFFER FOR SALE PAPER OF THE suspended houses of New York, Boston and Philadelphia. All orders promptly executed.   no25-1m:e47

SPERM OIL! SPERM OIL!
PATENT SPERM OIL.
AN EXCELLENT ARTICLE OF BURNING OIL for sale low by
J. H. REED & CO.,
no25   144 & 146 Lake street.

BOOKS AT REDUCED PRICES.

Advertisement for sperm oil. *Chicago Daily Tribune.* November 26, 1857.

of its Luster, the honest Enquires may be brought within her Boarders and become co-inhabitants and join with her." Irish minister Richard Shackleton urged Friends to remember that "This life is the field of battle…. May the lamp of God in the temple of our hearts, be kept still renewed and replenished with heavenly oil."[27]

Lit candles, burning lamps, heavenly oils—these were the actual, physical products that New Bedford produced every day from the maritime labors of the city's whaling sons. Odes were written to them and they were woven into civic celebration: "Glad youth and snowy age; Come on their pilgrimage to honor thee; Where burns thy lamp so bright; All hail thee with delight." These were just some of the illumination-centric lyrics sung at New Bedford's public semi-centennial celebration.[28] And the fact that the whaling industry's most prominent names were also on the rosters of one of two Quaker meetinghouses reveals just how intertwined the metaphors of lighting the world really were. Howland. Rotch. Rodman. Taber. These were the who's who of New Bedford's ruling class and New Bedford's whaling industry, and they were all Quaker families.[29]

Philip Hoare's 2010 meditation on whales reminds us that sperm whales— that most-pursued path to lighting the world—are themselves global animals, who "live in every latitude and every ocean, from the North Atlantic to the

South Pacific, even in the Mediterranean."[30] Thus, the men of New Bedford found themselves in every corner of the globe, in a cycle of pursuit, production, and consumption that linked the whale to the whaler, but then again between the whaler and the mythical bearers of light. The dedication to light and the quasi-religious concept of spreading light to the world was fully on display when the city was formally incorporated in 1847. An editorial in the *New Bedford Mercury* hailed the event with these words: "To the latest generation may she be a burning and a shining light! May she be illuminated with the oil of gladness and blessed with plenty and prosperity!"[31] It is not a difficult leap to assume that the "oil of gladness" was none other than the very whale oil and sperm oil the city produced to illuminate every aspect of American society.

Light was equated to both the prosperity and moral fiber of New Bedford. In Ephraim Peabody's 1842 sermon which took place at the commissioning of the New Bedford Orphan's Home, Peabody heralded the prosperous town where lights "blazed along the business streets," thus evoking the public sphere of men, commerce, and industry. But he also singled out the importance of illumination in the domestic sphere of women, where "in private dwellings, happy and light-hearted circles meet, *and under the glowing light*, amidst joyous words and music and song, the evening hours fleet on swift wings away [emphasis added]."[32]

James Almy's seal was a codification of all of this. The phrase *Lucem Diffundo* had now become an important launching point for New Bedford's sense of identity, purpose, and destiny. The switch from the more passive "spreading light" of *Lucem Diffundens* to the more personal and active "we spread light" of *Lucem Diffundo* helped telegraph this to the rest of the world. Whaling was more than an industry, and more than a pursuit of profits. There was a Quaker-influenced religious sheen to all those oil lamps and smokeless spermaceti candles. Men had been trying to roll back the impact of night for millennia. Now a city—literally a city upon a hill as James Almy's revised seal clearly showed—was accomplishing something that was nearly divine in scope. New Bedford was bringing light to the world and repelling the darkness.

Three

Somewhere in the South Atlantic, 1858

At least Eliza Williams had overcome her seasickness. When the *Florida* sailed from New Bedford in the autumn of 1858, she left with a crew of 37, plus Eliza Williams, the wife of Captain Thomas W. Williams. Eliza's presence on a whaling voyage might seem a surprising addition to our modern eyes; however, it was a much more common practice than we may realize today. The custom of women sailing with their husbands on these multi-year voyages had been gaining traction for some time in the years prior to the *Florida*'s departure. The *Whaleman's Shipping List* reported in 1853 that women accompanied their captain-husbands on about one in six voyages, thus preserving "the ties of domestic life that otherwise would be sundered." The definitive trade publication of the whaling industry also noted that these accompanying wives were "capital correspondents, and through the female love of letter-writing, keep us well posted up in the catch and the prospects of the season."[1]

Eliza Williams and the *Florida* would not return to American soil until October 1861, when the whaler anchored in San Francisco Bay. In the just over three years at sea the *Florida* would cover the Atlantic, the Pacific, and the Indian Oceans. Having taken both a blackfish (a pilot whale, often caught early in voyages and seen as a good "practice" opportunity for green hands or first-time whalers) and a sperm whale calf, it was not until early November in the South Atlantic that the crew would successfully kill and harvest their first adult sperm whale.[2]

In the meantime, Eliza experienced a range of thoughts and sensations that she recorded in her journal—seasickness foremost among them, at least at first. "It is quite rugged today, and I have been quite sick," she wrote only a few days into the voyage, "these 3 or 4 words I write in bed." "All I can think of is perpetual motion on board this Ship," she bemoaned four days after that. Eventually both the weather and Eliza's sea legs cooperated enough to get her mobile again, and soon she was able to record more than just the familiar

The Last Voyage of the Whaling Bark *Progress*

lament of a first-time sea traveler. The rest of her journal is a testament to the realities and truths of her husband's chosen profession, and her words provide a useful window into the dynamics of whaling. It was an intricate process that not only has been largely forgotten today, but often passed unknown or unacknowledged even when whaling was one of the United States' most important industries.[3] Eliza Williams thus becomes an almost Virgil-like guide into the many layers of whaling's hellish process.

Whaling was gruesome, dangerous work filled with life-or-death moments of violence, stamina, adrenaline, and heart-pounding courage. Whaling was painfully boring with months of tedium and nothingness that eventually accumulated into voyages that stretched on for years. Though seemingly contradictory, both statements accurately describe the life of whalers and their time aboard whaling ships. Days, weeks, and months would pass aboard a ship about 100 to 120 feet long … roughly the same length as a modern-day Boeing 737. Crews on the larger whaling ships and barks that were more common in New Bedford by mid-century hovered between 30 and 40 men—men in whose company you would expect to spend seemingly endless months living and working.

The American whaling industry was based on a long, complex, and interconnected chain of events, jobs, and tasks. The process of whaling began with a constant watch of the horizon from the masthead or highest part of the mast. From the masthead a crewman would stand for two-hour shifts. Usually the lookout location was simply "cross-trees," two pieces of wood nailed to the mast that served as a perch. Sometimes the lookout had a "masthead hoop"—a bit of rope or barrel hoop that encircled the man. Only Arctic voyages tended to have crow's nests, the more substantial man-sized buckets or even tents that would provide the lookout some decent protection from the elements. Melville describes serving as a lookout as an almost meditative (perhaps drug-influenced) experience in *Moby-Dick*. His Ishmael was "lulled into such an opium-like listlessness of vacant, unconscious reverie" by the ocean's waves.[4]

Of course, the lookout was there for a reason, and that reason became manifest to everyone on board when the cry "there blows" echoed from on high. Variations of "there blows" included the more famous "there she blows" which Melville used in *Moby-Dick*. The call was long, drawn out, and resonant so that it would carry over the entire ship. Many authors of whaling attempted to capture the effect by adding an inordinate amount of "o"s to the phrase, as in one account's "she blooooows, bloooow, bloows."[5] On the *Florida*, the lookouts used the "there blows" variant according to Eliza Williams, who also mercifully refrained from adding unwarranted extra vowels

Three. Somewhere in the South Atlantic, 1858

in her retelling. When the lookout's "there blows" spread over the whaleship like an audible fog on November 8, 1858, the ship had already been at sea two months. In that time the shout from above had only gone out three other times.

Once the cry occurred and the sighting of whales was confirmed by the captain using a spyglass—and assuming the whales were close enough to realistically be hunted—the whaleboats would be lowered into the water to give chase. It was these 30-foot, six-man whaleboats that pursued whales, not the fully rigged ship or bark. Whaleboats were the key to whaling and the cornerstone of the entire industry. Each of the *Florida*'s four mates commanded his own whaleboat and crew, as was the custom across the whaling fleet. Selecting his men would have been among each mate's first tasks at the start of a new voyage. Well-trained crews could lower a whaleboat from her davits in less than a minute. Powered by five ash wood oars of varying length, the whaleboats would need to cover the space between the ship and the whales as quickly as possible, but not so recklessly or noisily as to spook the whale into fleeing. And of course, the whale or whales did not just lay still. They too were in motion, often moving in the opposite direction from the arriving boats. This is why the whaleboats were so ingenious. Light and fast in a way the larger whaleship never could be, the whaleboats were designed for a chase across great expanses of ocean.

While five crewmen rowed at a backbreaking pace, one of the ship's mates (or sometimes the captain himself) would steer using a long steering oar. Often a small sail was also employed, in part to help minimize the noise of the whaleboat as it approached its target. In fact, according to her journal, one of the first things Eliza did when she started feeling better was to help sew a new whaleboat sail. But as the whaleboats departed from the *Florida* on a bright, calm November morning in pursuit of sperm whales, Eliza Williams noted the boats "were lowered and pulled lustily for them [the whales]," suggesting the strong physical labor needed for this part of the chase.

Anyone who has ever been on a modern-day whale watching excursion will certainly understand the difficult part that came next: actually finding the whales in order to get close to them. All whales dive underwater for long periods of time before they surface, and in that time, they can move significant distances from where they previously surfaced. A sperm whale can remain underwater for as long as 90 minutes. And given that they can travel as fast as 23 miles per hour under duress, the difficulty of being near to where they resurface becomes apparent. The skeleton crew left back on the whaleship would do what it could to help. Writes Eliza Williams of that November 1858 hunt: "On board the Ship, they place signals to mast head in different places,

The Last Voyage of the Whaling Bark *Progress*

and different shaped ones, made from blue and white cloth, to let those in the boats know in what direction the whales are and whether they are up or down, as it is difficult sometimes for the Men in the boats to tell, they are so low on the water and the whales change their position so often."[6]

Still, there were things that worked in whalers' favor. Decades of professional knowledge about animal behavior had been gathered by whalers and passed down through the ranks to each rising generation. Whalers knew that different behaviors—feeding, playing, courtship—would produce different likelihoods of direction and time underwater. Also helpful was that sperm whales tended to travel in a straight line, so much so that one whaler remarked, "It has puzzled many men as well as myself to account for their being able to keep a straight course for hours and never vary a point by the compass from it."[7] This obviously made it easier to guess the direction, if not the exact location of the next spout. Additionally, when sperm whales resurfaced, they then remained on the surface for several minutes to re-oxygenate, giving the pursuing whaleboats their opportunity.

"Sperm Whaling, No. 1–The Chase," 1859. Lithograph from drawings by Albert Van Beest and Robert Swain Gifford. The two whaleboats in this image include one using a sail and another relying solely on hand rowing. The whaleships from which they launched remain in the background. Although two whaleboats have arrived in close proximity to the surfaced whale, only one would have been allowed the attempt to harpoon the intended quarry (courtesy New Bedford Whaling Museum).

Three. Somewhere in the South Atlantic, 1858

Up until this point, the work described was difficult and strenuous, but perhaps not disproportionally dangerous compared to other maritime pursuits, save those brought on by rough seas or unexpected mishaps when lowering the whaleboats. That all changed once a whaleboat maneuvered close to the whale, as the *Florida*'s first mate Samuel Morgan did that November afternoon. Mates would steer whaleboats, but at this moment it was the harpooner, always the forwardmost oarsman in the boat, who likely had all eyes fixed upon him. He would stow his oar, working to keep as quiet as possible in order to maintain the element of surprise over the whale, and would then pick up one of the two harpoons that were kept at the ready. The harpoon was the most recognizable instrument of whale hunting, and core to the entire endeavor. The *Florida* had been outfitted with 200 of these "iron poles" upon leaving New Bedford, a reference not to the metal at the tip but to the six-foot long poles made from heavy hardwoods like hickory and oak. The harpoon's head was produced by a blacksmith, and by the mid–1850s a "toggle iron" design for harpoons had spread across most of the whaling fleet. They were sometimes referred to by the name "Temple toggle iron," in honor of its inventor Lewis Temple, an African American blacksmith and former slave who lived in New Bedford.

The harpoon was designed to be thrust deep into the whale's flesh and blubber once the whaleboat had maneuvered into position. It operated through the brute force of the harpooner, who braced himself against the "clumsy cleat"—a notch in the whaleboat—as he prepared to deliver the strike. He listened for the mate's command of "give it to him," indicating the boat was in position. Writes photographer and filmmaker James McGuane, "The ideal striking distance was 'wood to black skin'—virtually 'beaching' themselves on the whale's back. The iron was never 'thrown' but 'planted' or 'darted.' The word 'thrown' would seem to imply a certain lack of resolve—as if the men lacked the courage to get close and fasten to the beast properly." It was a tricky calculus—the closer the whaleboat got to the whale, the more the harpooner could thrust instead of throw, thus increasing accuracy and depth; too close, and the whaleboat risked being smashed to pieces once the harpoon struck and the whale "lashed out with unimaginable agility."[8]

As the harpoon slipped deep into the whale's flesh, with both flues, or barbs, fully buried, the genius of Lewis Temple's design became manifest. A small wooden peg holding the rear barb in its closed position would snap when the whale began to flee. Thus freed, what had once looked to be a single, smooth spearhead now became T-shaped as the entire harpoon head swiveled. This had the effect of holding the tool's head deep in the whale and making it nearly impossible for it to be pulled back out. Typically, an additional harpoon or iron would be thrust into the whale if a second strike was feasible.

None of this was designed to kill the whale. As dangerous and physically brutal as this stage was, nothing thus far could have in any way slain a 30 to 60-ton animal. The harpooning of a whale was simply a means to an end—in this case, the literal end of a rope. The harpoons were not simply the instrument of choice in whale hunting; they were also anchors between the whale and the whaleboat. Harpoons were attached to coiled lines of hemp rope known as the "whaleline." When a whaleline was stretched out from their carefully hand-coiled loops inside a tub they measured anywhere from 1,200 to 1,800 feet in length. While one end of the whaleline began at the harpoon in the whale's back, the other was secured to the whaleboat.

However, it wasn't quite as simple as running a straight line between whale and boat. The rope would be wrapped twice around the loggerhead, the wooden bollard at the stern or *back* of the whaleboat which was used to slow the line as it was run out by the fast-fleeing whale. Therefore, this rapidly uncoiling rope had to run the entire length of the whaleboat. With a fleeing whale capable of reaching speeds of 20 miles per hour or more, the rope or whaleline became a deathtrap as it unfurled along the length of the boat, capable of slicing off limbs or hurling men in the sea. Melville described the experience of anyone unfortunate enough to be holding onto the line without gloves as "like holding an enemy's sharp two-edged sword by the blade, and that enemy all the time striving to wrest it out of your clutch." Many greenhorns lost flesh and blood as they learned the power of a live whaleline. The rope's velocity was so fast that sparks from contact with the loggerhead were common, as was "a hempen blue smoke" as the line played out.[9]

Eventually the rope would cease unspooling from its tub and go taut. Thus would begin a phase of the hunt colorfully known as the "Nantucket sleighride," where the injured and enraged whale's flight suddenly became the mechanism for pulling six men behind it in the whaleboat. The lengths of rope involved were typically enough to keep the submerging whale from taking the whaleboat underwater with it. Instead, the whaleboat would skim, skip, and slap across the waves of the surface propelled at speeds of up to 20 miles an hour by the whale.[10]

Eliza Williams, of course, was not along for the ride that first mate Samuel Morgan took in 1858 once they were attached to their sperm whale. But whaling's literature is filled with accounts of this dangerous and adrenaline-pumping stage of the hunt. Many emphasized the speed with which the sleighride took place: "To the man who has exhausted even the delight of the

Three. Somewhere in the South Atlantic, 1858

sixty mile an hour automobile there is an unlimited field. The chances are that if he once gets an opportunity to taste the unbridled and terrific pleasure of a 'Nantucket sleighride,' he will view his auto machine as a tame thing ever afterward."[11] More akin to staying atop a galloping horse for hours, few men in 1850s America would have known what it was like to physically travel this fast for such a sustained length of time. A Pony Express rider, by contrast, only averaged about ten miles an hour. For a young man seeking thrills in the mid–nineteenth century, the Nantucket sleighride was one of the fastest experiences available.

The sleighride also necessitated unwavering attention and relentless reaction. Far from just "going along for the ride," the whalers would have executed infinite maneuvers and decisions. A constantly-shifting whale in the lead position meant a constantly shifting whaleboat behind it. This required the crew to sway and redistribute their bodies to ensure an even keel and to prevent the boat from capsizing—all while minding the nearly-alive rope running down the length of the whaleboat. Any slack in the line would be taken up through the loggerhead, bringing whale and boat closer and closer together, yard by yard, foot by foot, inch by inch. There was a real danger that the whale would double back underwater, come up beneath or alongside its unwelcome hangers-on, and launch an attack. Even the whale's natural maneuvers could be deadly, such as if the fluke or tale came down on the attached crew.

A Nantucket sleighride could also produce another problem, given a whale's speed and ability to travel long distances, as was recounted in one whaler's remanences: "up came the whale with a loud snort and started to speed away to the south like an express train [writing in the 1930s, the author had a wider repertoire of speed metaphors]. In fifteen minutes we were out of sight of the other boats.... When on top of the water, it kept going at an awful speed.... Looking around the waste of water, no other boat was to be seen, not even the ship." Taken miles away from their shipmates across a random expanse of ocean, the whaleboat crew endured days adrift without food and water before finally being rescued by another whaleship.[12]

After potentially hours of sleigh-riding, two things eventually happened. One, the whale would exhaust itself from the chase, its wounds from the harpoon(s), and the exertion of hauling around a whaleboat and her crew. And two, the winding in of the whaleline would ultimately bring the boat within "striking distance"—quite literally in this case—of the whale. Sometime during this mayhem, often immediately after the harpoon strikes, yet another step in the intricate and multi-layered process occurred at the mate's command. In this step, the mate took over the front position, armed with a new instrument—the lance or killing iron. The man who had been the harpooner now earned his actual title of boatsteerer, as he moved to the back to take over where the mate had been sitting. Though these crewmen were both harpooner and boatsteerer, the latter was the title given to them in crew lists and manifests. Explained one whaler after dodging a dying whale: "I knew then why a harpooner is called a boatsteerer: harpooning a whale is nothing compared to steering the boat after he starts fighting."[13]

The dangerous switch in positions also gave the whaleship officer the kill position. At first this might seem slightly ceremonial, even theatrical. Was all this maneuvering and position-switching done just to give the man of highest rank the "honor" of dispatching a whale, much like a general finishing a kill on the battlefield or an important client being offered the kill shot on a hunting safari? In truth, this moment in the whale hunt was so extremely dangerous and such a critical part of the entire endeavor that it required the experience and skills that ships' officers had acquired from previous voyages and hunts—skills that translated into a duty and responsibility for killing the whale.

The lance was a vicious instrument. The *Florida* was outfitted with 100 of them, each for deadly purpose. A five- or six-foot tool affixed to another six feet of wooden handle, the lance was an oval blade sharpened along its every edge. When first mate Samuel Morgan took his position, lance in hand, he would have been prepared to stab deep into the sperm whale's flesh attempting to hit the heart or lungs. The lance's ovalish diamond shape allowed it to slide in and out quickly, so that the mate could strike multiple times in quick succession. The other five men aboard the boat would have been looking for the telltale sign that the whale's vital organs had been hit and severed—when the whale started spewing blood out of the blowhole known as "chimney's afire" or a "red flag."

All of this was incredibly dangerous. A whale did not just lie there, allowing itself to be stabbed multiple times, even if exhausted from the chase. The ensuing "flurry" of a multi-ton animal flailing, bucking, churning, biting, and smashing its body (and especially its tail) into whatever was nearby brought about the destruction of many whaleboats and took the lives of scores of whalers. When William Henry Giles Kingston, the English author of children's adventure books, turned his attentions early in his career to whaling in his popular 1851 *Peter the Whaler*, his scene of a flurry was particularly vivid: "'Never saw a whale in such a flurry,' said old David, into whose boat I was taken. For upwards of two minutes the flurry continued, we all the while looking on, and no one daring to approach it; at the same time a spout of

Three. Somewhere in the South Atlantic, 1858

"Sperm Whaling, No. 2–The Capture," 1859. Lithograph from drawings by Albert Van Beest and Robert Swain Gifford. Here the mate has taken his position in order to inflict the whale's deathblows via a lance. Note the whaleboat in the near background has suffered a different fate, as the whale's "flurry" sends the boat and men flying (courtesy of the New Bedford Whaling Museum).

blood and mucus and oil ascended into the air from its blow-holes, and sprinkled us all over."[14]

Samuel Morgan and his five crewmen killed an adult sperm whale on November 8, 1858, sixty-two days into the *Florida*'s voyage. However, it was only after all of these steps—spotting a whale, lowering the boats, giving chase, setting the harpoon, riding the sleighride, shortening the whaleline, lancing the whale, and surviving the final flurry—that such a success could officially be declared. That final moment, and the pronouncement of a completed cycle of dangerous labor, came when the whale rolled onto its side, and the ensuing cry of "fin out" went up as the whalers spotted the quarry's fin sticking out of the water. Eliza Williams reported that the first cry of "there blows" came before breakfast. When first mate Morgan finally "got fast" to their quarry, the captain "sent two boats to him to help him tow the whale to the Ship. As he was some ways off and there was no breeze to help him, it was long after dinner before they came alongside."[15] Hours had elapsed just to reach this point. Six men had successfully killed an adult sperm whale, and 18 men had just completed the arduous task of hand-rowing a multi-ton carcass (literally dead weight) back to the *Florida* through open ocean. Yet despite all this hard, physical labor, the work of the whalers and the whaling ship was just beginning.

Once the whale carcass was hauled to the ship and fastened alongside to keep the body from sinking to the bottom of the ocean, the "cutting in" process began. Williams had written a bit about this process in her journal earlier in the voyage when the sperm whale calf had been taken: "The Officers seems to understand exactly where to commence cutting and how to cut him all, and just where the joints are, they have handled so many. They first take the blubber off with spades with verry [sic] long handles; they are quite sharp, and they cut places and peal it off in great strips. It looks like very thick fat pork, it is quite white."[16]

Her mention of pork and cutting at certain joints helps us today to imagine this step for what it mirrored, at least in part—a butchering process, similar to any stockyard or butcher shop. There were of course several key differences, chief among them: the size of the animal involved, the pursuit of blubber rather than cuts of meat, and the fact that this butchering process occurred in rolling seas. The latter obstacle was partially overcome by building platforms off the whaling ship and over the carcass, so some work could be done from above. However, as Eliza Williams noted, often the men simply stood on the back of the floating whale carcass. Also, a butcher of a cow or pig has likely never had to do his or her job while keeping an eye out for sharks—a very real threat to those performing a butcher job in open seas. Writing of his 1880s voyage, one former harpooner remembered, "The bug-light is a hoop-iron basket and burns whale scraps. It gives a very good light. We hang it over the side above the whales so that we can see the sharks and cut them in half with a cutting spade.... There are always at least two men killing sharks when cutting-in has to be carried on through the night, but during the day the men working on the whale take care of the sharks."[17]

When the November 8 whale was safely secured alongside, Eliza Williams got into her husband's whaleboat in order to have a better view of what was happening. "They begin to cut a great strip. The hook is put through a hole that is cut in the end of this piece by the boarding knife [a double-edged sword-like blade attached to a pole]. Then it is drawn up by the tackle as they cut. They do not stop till the piece goes clear around. Then it comes clear up...."[18] These fatty, slick strips of blubber called "blanket pieces" unwound off a whale (many have used the metaphor of a peeling an orange or lemon in a single strip) until they were up to twenty feet in length. This too was all exacting and specific labor, but the blanket pieces of blubber still were not the endgame of whalers. After all, the flesh would have rotted away in

Three. Somewhere in the South Atlantic, 1858

days, particularly in the warm climes in which the *Florida* found herself. The blanket strips therefore begat yet another process known as trying out. It was the process by which whale oil—the true end goal of the voyage and the captain—came about.

Whale ships were essentially floating factories for producing whale oil and sperm oil. The strips of blubber were the raw materials, which were handed off to crew members working in the "blubber room." Ishmael's description in *Moby-Dick* gives some sense of this space: "Down goes the first strip through the main hatchway right beneath, into an unfurnished parlor called the blubber-room. Into this twilight apartment sundry nimble hands keep coiling away the long blanket-piece as if it were a great live mass of plaited serpents."[19] The blubber room was the first stop in a multi-step chain to take the giant "plaited serpents" of the blanket pieces, which were nearly four times as long as a sailor was tall, and reduce them into page-like sheets, ¼-inch thick and a foot long. Hand-wielded mincing knives or a crank-driven mincing machine (which the *Florida* had) would be used to break the strips down, as chunks were passed back onto deck for further reductions.

On deck was another space designed to carry out yet another specific task. The tryworks were dedicated to boiling the blubber in large iron cauldrons called trypots. Once the blubber was cut down to pieces the size of a bible page (they were actually called "bible leaves") they would go into the cauldrons which were kept scorching hot by the fires beneath them. The entire kit of the tryworks often reminds modern-day viewers of a brick backyard barbecue.

These tryworks were where the coveted sperm oil and whale oil were finally seen for the first time, excreted into the searing bowl from burning flesh. Like a sort of maritime alchemy, the fires of the tryworks commenced the process of a solid being converted into liquid. The burning and searing continued until the flesh was crisped through and most of the oil released. Even these flesh scraps were recycled, pressed one last time for final drops of oil and then thrown into the tryworks fire as fuel. In addition to the blasting heat, the process had another hellish byproduct—smoke. As whale blubber rendered and liquefied into oil it released a thick, greasy, foul-smelling smoke that coated everything and everyone. It was the all-consuming, inky smoke belching forth from the tryworks that gave whaleships their reputations for being smell-able miles away. Eliza Williams put it mildly when she wrote, "The smell of the oil is quite offensive to me."[20]

So long as there were strips to harvest, mince, and try out, the crew

The Last Voyage of the Whaling Bark *Progress*

Untitled book illustration 1835–1850, after Ambroise Louis Garneray's "Peche de la Baleine" (Paris, 1835). This image shows the cutting in of whale blubber into long strips and the trying out process to render the whale oil. The infamous thick black smoke from this stage of the work covers much of the ship and surrounding area. Some wags from the era noted you could smell a whaleship before you could see one (courtesy of the New Bedford Whaling Museum).

worked non-stop, day and night. "Trying watches were six hours on and six hours off," writes one historian of the trade. "Continuous processing at a rate of 50 barrels a day could consume even the largest whale in two or three days...."[21] Over the course of November 9 and 10, the *Florida* fell in with a whole pod of whales and an additional four carcasses were affixed alongside her to work through at a dizzying pace. "The Men worked away at them till night and all night too," Williams wrote. "They don't stop cutting or trying as long as they have a bit of blubber on the Ship; they never let the fire go out till they are through."[22] Once rendered down, the whale oil would be poured into casks, the production and maintenance of which fell to another member of the crew—the cooper. During times of abundance, such as having five whales in three days to process, the cutting in of blubber could outpace its rendering into oil. Williams writes that the men in the blubber room "were at work up to their wastes in blubber."[23] Eventually the two sides of production—cutting in and trying out—would align and the

Three. Somewhere in the South Atlantic, 1858

final product of whale or sperm oil would be stored in the hold after it had sufficiently cooled.

Moreover, as all this activity and work was carried out there was yet another process—this one dedicated to collecting spermaceti from the head. Unlike the process involving strips of blubber, the spermaceti-collecting process occurred on deck. The head of the animal was severed and hoisted aboard ship. From there the "milk pail"-like process of scooping out the milky substance took place. On November 10 Eliza Williams woke to three whale heads on the deck and a fourth about to be hoisted aboard to join them. These too would have to be worked through, eventually yielding their prizes for storage in the hold.

Then there were still other products to retrieve from sperm whales if possible, particularly ambergris. Dr. Sturgeon Stewart perhaps best captured the ironic nature of this substance when he wrote a 1909 piece entitled "The Whale and his Haunts": "The 'ambergris' obtained from the sperm whale and which is so highly prized, and used largely for manufacturing fine perfumes, and in France and Turkey for aphrodisiac purposes, is nothing more or less than the retained anal secretion of a deceased sperm whale."[24] Indeed, the dark, waxy substance appears to be a product of the sperm whale's proclivity for consuming giant squid, and the whale's inability to digest the squid's sharp beak. The ensuing disruption concentrates in the whale's digestive tract until the resulting substance of ambergris can reach several hundred pounds in weight. To nineteenth century consumers, the somewhat disgusting end product nevertheless enabled perfumiers to manufacture concoctions that held their scent and lingered in a sort of stasis rather than evaporate. Given the rarity of ambergris, a successful retrieval from a sperm whale's intestinal tract could literally generate half the profits for some of the shorter voyages.[25] Sperm whale teeth were also removed and collected, typically as souvenirs or to carve into scrimshaw pieces.

Even after all the aforementioned time and work, the process did not let up. Cleanup came next. Removing the gallons of oil, blood, and guts that coated the decks was the necessary last step. As one whaling account put it, when the "laborious business" of trying out was completed, the work of cleanup began: "the tubs, knives, &c., were removed below, and the ship received a thorough scouring fore and aft, with strong alkali and sand applied with the scrub brooms. And indeed she required it, for the muddy scurf from the exterior of the whale uniting with the oil, does not improve the appearance of anything with which it comes in contact."[26] Like everything else in the process, this was done in a thorough and complete fashion—no shortcuts would be allowed by the officers in charge. In the words of another whaler, "when all was done you would have found it hard work to soil a white pocket handkerchief by rubbing it on any part."[27] Given the level of gore that the process produced aboard a whaleship, this was quite an accomplishment.

All of this labor, all of these steps, was the proscribed and regimented way that whaling took place. Each one brought the crew of a whaleship closer to a full hull of oil and closer to the captain's eventual decision to finally head home. Each was a link in the chain leading to the fortunes that whalers were seeking in every ocean and across every corner of the globe. None could be skipped. None could be accomplished through shortcuts. And they cycled over and over again, beginning with each whale sighting and the call of "there blows" and ending with a freshly scrubbed deck. The repeating wheel of labor turned to each task, filling the whalers' days with excitement, intense work, and exhaustion. Except when they didn't.

After taking whales for two days in November 1858, the crew of the *Florida* spent another furious four days going through the trying out, storage, and cleaning process. The deck heaved with casks waiting to be taken to the hold, and men congratulated themselves on the fortune of product they had captured and rendered. The luck continued through the rest of year, with several more captures before the end of 1858. And then ... nothing. Eliza William's journal shows that the crew of the *Florida* did not capture a whale again until July 16, 1859. Months had passed with occasional sightings and chases; however, they had always ended with the crew returning to the ship empty-handed. This too was the reality of whaling. It often stretched on for months without fulfilling its ultimate goal. This is why whaling voyages came to last for years—it took that long for the journey to be profitable when the space between the production of whale and sperm oil could be marked in seasons, not days.

It is safe to say that few people outside of whaling communities like New Bedford or Nantucket really understood any of this. Certainly, popular periodicals described whaling for interested audiences from time to time, but whaling was laden with tasks and terminology that were largely unintelligible to lay audiences. Nobody understood, much less cared about, "hawseholes" (the holes through which anchor cable passed), "drogues" (objects of additional drag added to the whale line to slow down the fleeing whale), "piggins" (scoops used to bail whaleboats), or the hundred other words uniquely understood within the fraternity of whalers. *Moby Dick* famously remained mostly unread by American audiences until well after Melville's death in 1891, as did another realistic portrayal of the brutality of whaling—Edgar Allan Poe's *The Narrative of Arthur Gordon Pym of Nantucket*.

Three. Somewhere in the South Atlantic, 1858

What American consumers mostly understood was a sanitized fantasy of whaling that emphasized heroism and bravery. Whale oil products invariably showed whaleboats lowered and harpooners in thrilling action rather than the grimy floating factory process of cutting in and trying out. City boys might have seen whaling "representations" or lectures that toured urban theaters and halls, but these inevitably emphasized "monsters of the deep" and exotic native women over the technical details of whaling. Writes one historian of the genre, "they helped lure scores of landlocked young men to sea, where months of drudgery convinced many that they had been hoodwinked."[28]

Popular dime novels sprinkled in whaling terminology for effect, but likewise disproportionally gave readers tales of otherworldly savages, dastardly mutineers, and dramatic rescues. A word search of the 1865 dime novel *The Golden Harpoon or Lost among the Floes: A Story of the Whaling Grounds*, for example, reveals that the word "blubber" does not appear once in the nearly 100 pages![29] One of the most popular stories, the aforementioned William Henry Giles Kingston's *Peter the Whaler*, excised much of the grimness that defined whaling in favor of death-defying adventures and simple homilies about the virtues of "going a-whaling." Asked by his father about coming back poorer from his whaling voyage than when he left, Peter replies: "No, father.... I have come back infinitely richer. I have learned to fear God, to worship Him in His works, and to trust to His infinite mercy. I have also learned to know myself, and to take advice and counsel from my superiors in wisdom and goodness." Like so many dime novel authors, Kingston's goals were to provide a moral compass to his young readers more than a taxonomy of an industry. Replies Peter's father to his son's little speech: "Then I am indeed content, and I trust others may take a merciful lesson from the adventures of PETER THE WHALER."[30]

Thus, the detailed knowledge about how whaling worked and all the varied steps in a grueling and exhaustive process remained insular to industry itself. In part, this is what allowed whaling to flourish as an occupation for young men. In the absence of concrete information and widely-circulated knowledge about its inner-workings, whaling was able to project an image of adventure, freedom, and bravery. Future greenhands simply knew that they would be hunting nature's largest creatures. The steps in the process remained opaque and inconsequential to such narratives.

Yet even those with no intention of signing up for a whaling voyage were impacted by the dearth of knowledge about the process or the industry. As the complexity of whaling was stripped out, it left most Americans blissfully ignorant of where their whale oil and spermaceti candles came from. But this

The Last Voyage of the Whaling Bark *Progress*

Three. Somewhere in the South Atlantic, 1858

also meant they had no emotional or empathetic connection to the actual work or labor, and perhaps more importantly, to the actual workers themselves. New Bedford may have lit the world in the mid–1800s, but few outside New Bedford knew how that light was produced. Which begged the question: What if things changed? If the conditions and sources of light were to somehow be altered across the American landscape, would all those millions of people currently warmed by the light of whale products have any reason to care?

Opposite: Half Title, *Peter the Whaler* by William H. G. Kingston (New York: C. S. Francis & Co., 1852). Like much mid-century popular culture, the widely read *Peter the Whaler* trafficked in simple narratives and pleasing images of whaling. Inside, the reader found exciting descriptions of the whaling process, but they were almost totally devoid of actual whaling jargon or terminology. The half-title's inclusion of polar bears, tridents, and perhaps the scales of a sea monster also suggests broader adventures than the monotony of a multi-year whaling voyage (Baldwin Library of Historical Children's Literature, Special and Area Studies Collections, George A. Smathers Library, University of Florida, Gainesville, Florida).

Four

New Bedford, 1865

David Kempton was in a jam. The 47-year-old had faced adversity before, but circumstances over the past few years had particularly backed him into a corner. Orphaned at the age of twelve, Kempton left school and learned carpentry from the uncle who had taken him in. He struck out on his own afterwards, plying his skills with enough success to build up a nest egg—a sum of money he could use to join the booming whaling business that surrounded him. By age 28 he began investing in whaling voyages.[1]

Each whaling voyage was a discrete, individual opportunity for the citizens of New Bedford to invest. Some of New Bedford's wealthiest families owed their success to large investments in whaling. Sailmakers, rope makers, coopers, and other professional artisans were also commonly among those listed as part of a voyage's "ownership group," i.e., shareholders for a particular voyage. But a wide and diverse swath of New Bedford citizens also dotted those "owner" lists—bakers, jewelers, doctors, politicians, and ministers to name just a few. Though he was describing Nantucket, Melville slyly captured this system in *Moby-Dick*, writing: "the other shares, as is sometimes the case in these ports, being held by a crowd of old annuitants; widows, fatherless children, and chancery wards; each owning about the value of a timber head, or a foot of plank, or a nail or two in the ship."[2] So when a young David B. Kempton began putting his homebuilding profits into shares of whaling voyages in the 1840s, he found himself in good, varied company.[3]

Whaling in the world of David Kempton's youth was a smart, if fraught, investment opportunity. Rates of return—assuming a whaling ship didn't sink in a hurricane or become crushed in ice—were reliably better than agricultural investments, and often better than investing in manufacturing or transportation enterprises. A member of a voyage's "ownership group" might recoup profits of 6.5 to 14 percent, which was the typical pure profit rate when a ship finally returned to port. But for those who fell in love with not only the financial opportunities, but also the mystique of whaling, simply joining fellow New Bedford citizens in an investment pool was less than satisfactory.

Four. New Bedford, 1865

For David Kempton, the next step was to become a whaling agent, described as "the moving spirit of the industry."[4]

Whaling agents might find modern kinship with today's movie producers. The task of moving a complex, multifaceted, multi-staged enterprise from beginning to end became an increasingly specialized occupation within the whaling world, just as it later would in Hollywood's filmmaking universe. An agent typically (though not always) owned the physical whaling ship and would oversee its management. He was usually the largest shareholder within an ownership group, and often oversaw the investment process that doled out the remaining shares. He outfitted the ship for the voyage, chose the captain, and contracted firms to supply the crew. He arranged insurance, provisions, sails, rigging, and equipment. He also was responsible for selling off all the product that a captain and crew produced, keeping careful accounting records of the product in and product out, and the expenses incurred at each step.

With all this responsibility came the opportunity for great wealth, and agents were some of New Bedford's wealthiest citizens. These were the men and families that David Kempton knew of and admired, and perhaps even had worked for as a carpenter. Now, having built up his earnings through investing in whaling, he wanted to take the next step of purchasing a ship and becoming her agent. In 1850 David Kempton was ready to enter this arena by taking over his first whaler, the *Ontario*, which he bought and brought up from Sag Harbor, New York. He named his firm the David B. Kempton agency, and designed his agent's flag to fly over his new purchase—a white cross on a red field, vaguely reminiscence of the Danish flag. It was an auspicious beginning. Over the next few years, Kempton would become the managing agent of several more whalers including the *Waverly*, the *Kensington*, the *Christopher Mitchell*, and the *Barnstable*. When a display of agent flags and ships was published by New Bedford's Charles Tabor & Co. in 1857, Kempton's name was right there, nestled among some of the most important and prestigious family names in town.[5]

By January 18, 1861, Kempton had three whaling ships at sea, plying the world's oceans, and another ship in harbor at New Bedford. As it happened, on that same day, another hugely successful New Bedford agent named Matthew Howland sat down in his grand home to write a letter to one of his captains who was also far from America's shores in search of the industry's maritime prey. "We are in the midst of great excitement at this time," wrote Howland, "as South Carolina indeed the whole 'South' is threatening to seceed [sic] from the 'Union' from the fact of our having chosen a Republican President. What the result will be we know not, but already it has produced great stagnation in trade and business of all kinds."[6] In fact, the results would be much worse than Howland might imagine, especially for David Kempton.

It didn't start out that way. Instead, 1861 and the start of the Civil War brought opportunity for Kempton. All of Kempton's ships were relatively old. As a young agent starting a firm from scratch just a few years earlier, he had necessarily purchased older, less expensive vessels. The *Waverly* had been launched in the 1830s—three decades earlier—as had the *Christopher Mitchell*. Like many agents, Kempton likely wondered how many more voyages his ships could endure without major repairs or worse, outright loss.

All of which is why the start of the Civil War surprisingly offered a silver lining for Kempton and others, in the form of a strange offer which blew into New Bedford alongside the autumn chill. In October 1861, Gideon Welles, Secretary of the Navy for the Union, began issuing orders to fulfill President Lincoln's desire to blockade Southern ports, especially Savannah and Charleston. Welles' assistant secretary Gustavus Fox thought that such a blockade would be possible by sinking vessels in the channels of each port. So, in October, Welles instructed a purchasing agent in New York to begin secretly buying up 25 ships of at least 250 tons each, with the express intent of sinking them. Of the 25 vessels purchased under Welles' orders, 24 were whalers and 16 came from New Bedford. It is in this strange request that Kempton saw an opportunity and sold the *Kensington* to the United States government for $4,000— about $100,000 in today's dollars.[7] A few days before Thanksgiving 1861, the first of two "Stone Fleets" set sail from New Bedford. They were so dubbed for the cargo of granite blocks, boulders, sections of wall, and even cobblestones they carried. All the rocky variants were onboard so that the ships would sink quickly when scuttled. Kempton's *Kensington* was among them.[8]

The same year Kempton also sold off the *Christopher Mitchell* to a buyer in San Francisco who would convert her for use in the merchant trade. In 1864 he sold off the *Barnstable*, also to be fitted out for commercial use out of New York. None of this was particularly unusual. By most accounts, agents were more than happy to shed aging vessels during the Civil War, and many used the opportunity to consider exiting the industry entirely. Families that had once been synonymous with whaling began diversifying their investments. New Bedford dollars found their way to manufacturing, real estate, banking, and even the newly emerging oil refineries that ironically would come to compete with whale oil for consumers' lighting needs. Kempton's fellow agent Matthew Howland wrote in 1862, "a large number of ships have left the whaling business. I think about 100 in 2 years and probably a considerable many to arrive this year and now in port, will never be fitted again...."[9]

As David Kempton worked through his various networks and connec-

Four. New Bedford, 1865

"View of the Stone Fleet Which Sailed from New Bedford Nov. 16th. 1861," 1862. Lithograph from painting by Benjamin Russell. Multiple agents in New Bedford sold off older ships to the United States government to be used in the "Stone Fleets." David Kempton's aging *Kensington* is the last foregrounded ship on the right (courtesy of the New Bedford Whaling Museum).

tions to sell these ships, it is worth considering his overall position in New Bedford. Ultimately, Kempton was a relatively minor agent within the whaling community, and by the 1860s he was still nowhere near the status of the local industry's scions who he had perhaps admired as a child. By one historian's analysis, he ranked number 42 out of 43 among New Bedford whaling agents in terms of his firm's peak value.[10] Others in his shoes might have "taken the money and run" towards other industries or investment opportunities, as was already beginning to happen even within the relatively insular community of whalers. Kempton did not do this. And by December 1865 that decision had put smalltime whaling agent David Kempton in a very uncomfortable position.

Instead, he had plowed his capital back into buying whaling ships. In 1863 he bought a brig named *Leonidas* in Westport and converted her into a whaling bark. He also purchased the bark *Samuel and Thomas* from Mattapoisett. In 1864 he bought a whaler from Nantucket called the *Spartan*. And in 1865 he bought another Nantucket bark named *Islander*. In three years he had completely turned over the whaling vessel inventory under his ownership, except for the *Waverly*, which he still owned and managed.

The Stone Fleet ultimately failed. Blockade runners simply found alternate routes out of the harbors. But what it did accomplish was to remove

The Last Voyage of the Whaling Bark *Progress*

many ships from the New Bedford fleet, exacerbating the process that Matthew Howland noted in his letter. Moreover, it would be the later years of the Civil War that would deliver a heavier blow to the whaling industry. So far as anyone in New England knew in 1861, the Confederate Navy consisted of one warship, the *Sumter*. New Bedford, and indeed most of America, were not yet aware that a secret agent, James Dunwoody Bulloch, had been sent to Europe to find ships to be used as commerce raiders by the Confederate States of America. The South was determined to use their ragtag and motley collection of purchased ships to disrupt the North's economy. If in the process they also exacted revenge for the Stone Fleet—an act seen by the honor-obsessed South as despicably spiteful and dishonorable—well, so much the better. As a result of Bulloch's efforts, two Liverpool-built ships named *Alabama* and *Florida* began scouting the Atlantic in search of prey. The *Alabama* alone would take 70 Yankee ships of various sizes and descriptions until she was finally sunk in 1864.

For Kempton, these raids would produce the first of two disasters. In the early months of 1864 Kempton's *Samuel and Thomas* found herself in an enviable position—a run of good luck had brought enough whale captures and days of trying out to add several thousands of gallons of whale oil to her hold. But the ship was only a few months into its voyage—far too early to return to New Bedford. It was not an uncommon situation, and throughout the world, whaling captains worked through agents in various ports to have New Bedford-bound ships return with their cargo. Which is how it came to be that halfway around the world in Talcahuano, Chile, the *Samuel and Thomas*' captain arranged for 5,724 gallons of recently-rendered whale oil to be loaded onto the whaling bark *Golconda*, bound for New Bedford.

The *Golconda* set course, and Kempton's oil had made it as far as the coast of Bermuda by July 8, 1864. Unfortunately, that was when the *Golconda* encountered the Confederate raider *Florida*. Fresh off a few days of repairs in Bermuda, the *Florida* had already successfully captured and burned the *Harriet Stevens* just a few days earlier. Now the Confederates had the *Golconda* in their sights. The raiders took some oil for themselves, but ultimately set fire to the majority, along with the *Golconda* herself. In an instant, the whale oil that served as the lifeblood of Kempton's (and several other agents') enterprise went up in flames.[11]

Still, if a few months' worth of whale oil was his worst loss of the war, perhaps David Kempton considered himself lucky. Such cargo was undoubtedly insured, though at rates that had become exorbitant during the conflict. And as 1864 rolled over into 1865 it appeared that the bloody carnage of the Civil War would finally come to an end. When General Lee surrendered to

Four. New Bedford, 1865

General Grant at Appomattox Court House in April 1865, bells rang through New Bedford as they did through the rest of the North. Kempton could turn his attention to the future—which is why the events of that June must have been so jarring.

By June 1865 two months had passed since Appomattox. Although news of the surrender spread quickly throughout the states, the Bering Strait was a long way from where Lee handed over his sword and army—too far away, as it would turn out. Lieutenant Commander James Waddell, a lanky North Carolinian with thick sideburns, oversaw the newest rebel raider, the *Shenandoah*. After purchasing an old merchant ship for the express purpose of going after the North's whaling fleet, Waddell took his newly christened raider from the Madeiras Islands off the Moroccan coast of Africa in October 1864. It was a rocky start that ended with necessary repairs in Melbourne, Australia; however, from there the *Shenandoah* headed into the wide-flung Micronesian archipelago before finding four whaleships at anchor off Ascension Island, today's Pohnpei. The whaleships were quickly captured, looted, stripped, and destroyed.

Among the items Waddell took from the ships were charts showing where the Yankee whaling fleet would be concentrated during the summer months. Waddell plotted a course for the *Shenandoah*, heading to the Bering Strait. He attacked and burned whaling vessels as he went, thus preventing news of his approach. In an unfortunate twist of fate, several whalers of the fleet had converged in a single spot, coming to the aid of the *Brunswick* which had become trapped in ice. At that point, Waddell was able to corner and burn eight ships in one fell swoop. In total, the *Shenandoah* cut a huge swath through the fleet, capturing and burning 34 whaling ships and barks, and bonding (extorting a future ransom payment in exchange for safe release) another four. Most of them were from New Bedford. Kempton's *Waverly* was among the list of casualties, weighted down by 500 casks of oil that were used to torch the whaler while officers and crew watched as prisoners from the *Shenandoah*'s decks.[12]

Kempton had now lost not only two ships' worth of cargo, but a whaler as well. It was also the last of his original fleet of ships, dating from when he first became an agent. News of the *Shenandoah*'s assault filtered back in the summer of 1865. The *Evening Standard* wrote, "The blow inflicted on New Bedford by the destruction of her whalers during the rebellion and since its practical close is the severest that has been suffered by any community in the country, and the feeling of depression produced by the late news is much increased by the feeling that she has not had her fair share of protection from the government."[13] Kempton surely felt the burden of these losses weighing

The Last Voyage of the Whaling Bark *Progress*

"Destruction of Whaleships off Cape Thaddeus, Arctic Ocean, June 23, 1865, By Confederate Steamer Shenandoah," 1874. Lithograph from painting by Benjamin Russell. The image shows only a fraction of the whaleships that the *Shenandoah* destroyed, although the flames on the horizon suggest more destruction in the distance. David Kempton's *Waverly* was among the losses, though it is not depicted here (courtesy New Bedford Whaling Museum).

heavily upon his shoulders. Here was the dilemma in which the now 47-year-old agent found himself in 1865, when other parts of Massachusetts and the North were surely still celebrating the war's end and welcoming home war-weary soldiers. He needed to ensure that his remaining ships were refitted, crewed, and repaired as quickly as possible—and get them back out to sea doing the only thing that would make them profitable.

The *Leonidas* he turned around in barely a month's time—from when she arrived in port on August 18, 1865, to when she sailed again on September 21. The turnaround was remarkable, given the average number of days in port between voyages that year was approximately 215 days. It was certainly one of the year's fastest turnarounds, and it shows how eager Kempton was to maximize the resources he still had at his disposal.[14] His most recent purchase, the *Islander* (bought before or after news of the *Waverly* arrived is unknown), set sail on November 11. The *Samuel and Thomas* wouldn't return to port until 1866. That left one ship—the *Spartan*.

The *Spartan* returned to New Bedford on October 29. Even as work was underway to get the *Islander* out to sea, Kempton now doubled his efforts to

Four. New Bedford, 1865

also see the *Spartan* return to whaling before the winter settled in and closed New Bedford's narrow channel leading to open seas. But why had the *Spartan* returned at all? She had spent less than a year at sea since Kempton had purchased her from Nantucket in 1864. Her maiden voyage under Kempton's ownership had lasted a paltry 340 days. Although numerous explanations are possible, it seems likely that Kempton recalled the ship because he wished to replace the captain.

Agents picked the captains for each voyage, along with some key officers such as first and second mates, and other important positions like blacksmith and cooper. Agents and captains worked together on laying out the whaling itinerary before each departure, closely collaborating to maximize returns. Still, it would be a mistake to assume agents and captains always got along. Matthew Howland certainly could exhibit a biting tone when writing to his captains: "I regret that after your having decided to go the Arctic, you did not stick it out to the end of the season, oil or no oil..." Howland icily wrote in one example. "[Your letters] were duly received and their contents satisfactory, only we should have been glad if you could have reported 150 bls [barrels] sperm oil on board, but I have no doubt you tried hard..." went the text of another. Still a third suggested: "I am afraid I am doomed to disappointment, although I am aware we cannot always tell how a voyage is going to turn out, until it is ended."[15] If Howland's correspondence is indicative of the relations between agents and captains, then clearly such matters were not always smooth sailing.

Although the *Spartan*'s captain, Leonard B. Brownson, had previously worked with Kempton, serving as master of the *Barnstable* before Kempton sold her, it is notable that Brownson did not command a voyage again after the *Spartan*'s return that October in 1865. If Kempton did indeed recall his captain with the intention of replacing him, then Brownson's lack of another helm afterwards would suggest that word of the circumstances surrounding the dismissal had spread.

As work began to outfit the *Spartan* for another voyage, Kempton now had another job to add to his list—finding a new captain. He found Daniel W. Gifford, who by later accounts went by "Dan." The Gifford name was prominent in New Bedford. Giffords had been part of the New Bedford whaling community for over 40 years. Between 1860 and 1865 seven other Giffords besides Dan commanded New Bedford whaling vessels. William Gifford was one of the most powerful agents in town. His own ship, the *William Gifford*, had been captured and burned by the *Florida*.

But Dan Gifford was not from one of these more prominent Gifford families. His father was a carpenter, just as David Kempton had once been,

The Last Voyage of the Whaling Bark *Progress*

although he likely specialized in shipbuilding rather than homebuilding. The elder Gifford had died a decade earlier, leaving Dan's mother a widow with a brood of ten children. At least two of Dan's brothers were shipwrights like their father. For someone born into the trade, it could be a decent line of work. As the *Whalemen's Shipping List* reported in 1850, "SHIP BUILDING— We are glad to learn this important branch of 'home manufacture' has been revived in this vicinity with wonted vigor, and promises to afford full employment for a large class of industrious and enterprising mechanics and others."[16]

But Dan chose a life at sea, shipping out on his first voyage at the age of 16. According to the assorted shipboard records accumulated during his lifetime, he was 5'8" with a dark complexion, brown hair and brown eyes. Born in 1839, he would have been 26 years old as he discussed leading this expedition with David Kempton. Although not unprecedented, most men did not usually achieve the rank of captain until their thirties.[17] On the *Spartan*'s prior, shortened voyage, Gifford had been the first mate, the first time he had served on a vessel in that capacity. The journey before that he had served as third mate. Now Kempton was considering making him captain on his very next trip. With winter approaching, there was little time for deliberation. Dan became Captain Gifford and was given the *Spartan*.

While work to outfit the *Spartan* continued at a frenzied pace, Captain Gifford's life marked one additional milestone. On November 22, just under a month before he set out on his voyage, Daniel Gifford married Lucy A. Little, a 24-year-old woman from Dartmouth. Her parents were Esek and Ruth Little. They were married by clergyman Issac Dunham in New Bedford. In less than a month, Dan had become captain and husband. Now, as the days became shorter and the nights colder, he turned his attention to the sea.

The *Spartan* sailed on December 19, 1865, a mere 50 days after she had pulled into New

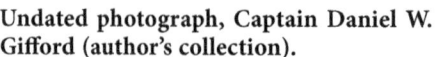
Undated photograph, Captain Daniel W. Gifford (author's collection).

Four. New Bedford, 1865

Bedford. The icy hand of winter closed in shortly thereafter. The *Spartan* was the last vessel to leave New Bedford that year. The next would be over four months later during the spring thaw. Kempton had managed to both replace his captain and return his ship to sea for a proper two-and-a-half-year voyage. He could breathe a sigh of relief.

David Kempton's business survived the Civil War, having undergone a rollercoaster of windfalls and setbacks. Given the opportunity to cash out of whaling in 1861, he instead doubled down on the industry, even as his peers diversified their investments or withdrew from whaling entirely. In the process, he had lost a season's worth of whale oil, as well as one of his ships, and replaced a captain with a young, largely untested substitute. All of this seems to beg a relatively simple question: Why? Although he certainly could not have anticipated any of these specific setbacks, he saw the same signs and heard the same prognostications as others in New Bedford that whaling was a declining industry. So why did he continue with the pursuit of whales not only through the Civil War, but onwards until 1877 before finally shuttering his whaling agent business and folding away his red and white owner's flag?

Perhaps he saw opportunity in the thinning of the field. Fewer ships might mean a smaller inventory of whale oil, which could then boost prices. His more established peer Matthew Howland made much the same argument in his letter lamenting the loss of active ships in 1862: "those [ships] now out will stand a chance to do well if they obtain some oil."[18] And it certainly is tempting to see an entrepreneur like Kempton, standing on the cusp of America's Gilded Age, looking at his future solely through the lens of dollars and cents. But to do so robs Kempton and others like him of much more human emotions and impulses.

In addition to the ledger books and anticipated profits, it seems equally important to consider that David Kempton was reluctant to give up on a dream. When a descendent presented her genealogical history of the Kempton family tree to the Old Dartmouth Historical Society in 1908 she noted, "No Kempton ever owned a wharf or had a ship named for him; for over a century after the family settled in Dartmouth, only one engaged in the whaling business, the late David B. Kempton."[19] Kempton wasn't born into whaling, he chose it—the only one in a century's worth of Kemptons to do so. Drawn to the industry from the margins of New Bedford's working class, Kempton seemed to crave the prestige that agents and their families commanded in the insular world of whaling. Looking at his choices from that perspective, the fact that Kempton hung onto his dream longer than was prudent or expedient would also suggest motivations of the heart, not just the head.

Some men would stick by whaling even after Kempton took down his shingle. They too harbored similar dreams and similar motivations. They balanced the ruthlessness of the forthcoming Gilded Age's devotion to capitalist profit with something much more esoteric and ethereal. They did so against the backdrop of a much bigger exodus away from barrels of sperm oil and risky multiyear voyages. For those that moved forward with whaling after the Civil War, their decision necessarily became much more conscious, much more deliberate, and ultimately, much more rooted in identity.

Whereas once almost everyone in New Bedford had been touched by and affiliated with the whaling industry, the David Kemptons and Matthew Howlands of the postbellum era were members of an ever-shrinking society. They were askew to and different from an ever-expanding new majority—those who no longer had connections to whaling. As with all forms of identity, this occupational and industrial connection came into greater relief precisely because of its growing rarity. David Kempton's decision in the 1860s to stick with whaling would convert him something unimaginable a generation ago—a minority.

Five

The Arctic, 1871

All Captain James Dowden could do now was wait. Twenty-four hours earlier he had been among a small group of whaling captains who had concluded a remarkable meeting near the top of the world, surrounded by steel-grey water, frigid air, and the ceaseless motion of gathering ice nearby. Seven captains in all, Dowden included, had just made the decision to give up the pursuit of Arctic whales for the season, and await the arrival of a very different quarry—a fleet of whaleboats carrying more than 1,200 whalers along with a few women and children. Over the next couple days, Dowden and the crew of the *Progress* would scan the horizon for this unusual flotilla to appear, anxiously wondering how their fellow New Bedforders were faring, as well as keeping an eye on the menacing ice and weather off the coast of the aptly-named Icy Cape. Even as the spouts of bowhead whales could be heard around them, the captains did not order boats lowered for pursuit. Instead, they waited. And, as James Dowden of the bark *Progress* told his peers, "I will wait for them as long as I have an anchor left or a spar to carry a sail."[1]

Dowden was one of the decreasing cadre of men who came out on the other side of the Civil War, the Stone Fleet, and the Confederate raiders, and continued to pursue whales. But much had changed. While whalers had been working in the Western Arctic and Bering Sea since the 1840s, the practice had steadily increased over the decades. For captains like Dowden in the 1870s, a trip north during the summer months had become an expectation. Agent Matthew Howland wrote to two of his captains in the early months of 1871: "I hope you will have better weather in the North this season" and "I am in hope you will have a better time for the season now coming." Neither was a request or suggestion about visiting the Arctic, only well wishes for what was clearly an assumed order about where the captains would be spending their summer months.[2]

The shift in focus to the Arctic was partly because the purpose and intent of whaling had also shifted in the interceding decades. Where sperm oil and whale oil had once reigned supreme as the ultimate goal of whaling, the demand for those products had been in a sustained decline. The discovery of petroleum in Pennsylvania in 1859 eventually brought new, cleaner-burning products to the market, especially kerosene distilled from petroleum. Coal-gas lighting in cities and across urban grids also had taken demand away from whale oils and spermaceti candles. Even New Bedford herself began laying gas pipelines to homes and businesses as early as February 1853.[3] This little bit of irony did not go unnoticed. "Is it not strange," asked the Seaman's Bethel's the Rev. Moses How, "that New Bedford which is emphatically the Oil Factory of the world, should abandon Oil which Is for their own interest to burn, and follow the example of other Cities in the Use of Gass? ... This is the age of improvement in almost every thing, but Religion."[4]

Fortunately for the whaling industry (and unfortunately for its prey), whales were able to offer their hunters an entirely different kind of product—baleen. Customarily called "bone" by whalers, baleen is plates of keratin (the same material that forms hooves and horns and fingernails in other mammals) that hang from the upper jaws of certain whales. Whereas sperm whales are toothed animals, species like blue whales, humpbacks, minke, and bowheads are all baleen whales, capable of growing 300–350 pairs of baleen plates over their lifetime. Whalers did not ignore the oil that toothless whales could produce, and tryworks continued to boil blubber and even whale tongues to extract that substance. But just as crews were accustomed to working the severed heads of sperm whales aboard the deck to scoop out buckets of spermaceti, now they hoisted up the severed heads of baleen whales for an entirely different product. The baleen plates—those from bowheads were often 10 or more feet in length—were cut away from the jaw and allowed to dry on deck before being bundled and stored below in the same hulls that had once bulged with barrels of oil.

The use of the baleen plates differed between whales and man. For the whales, baleen is a tool in the service of gathering and consuming food. The inner bristles of each plate form a fine, massive sieve through which ocean water passes. The sieve catches all manner of whale delicacies, but plankton especially, which are then swallowed in massive quantities with any remnants scooped up by that oil-rich, heavy tongue. But for humans, baleen plates were something else entirely—a light but strong, and most importantly, flexible material that could be cut, molded, and shaped into all manner of products. Shoe horns were made from baleen or "whalebone." So were eyeglass frames, umbrella ribs, brushes, canes, collar stays, and buggy whips. But all these paled in comparison to the one product that dominated the whalebone market. And in an ironic twist for a profession driven by men and masculine ideals of adventure and danger, it was a product that was entirely dependent on the fashions of women.

Five. The Arctic, 1871

Prior to the Civil War, the desire for larger and larger skirts was often supported—literally and figuratively—by whalebone. Many of the hoop skirts so beloved in images of the "moonlight and magnolias" version of the South needed something to sustain their ever-widening diameters. Whalebone was one widely available product that fit the bill. Thus, whalers in the 1840s and 1850s could maximize profits by simultaneously pursuing both toothed whales (for oil and the best products from sperm whales) and baleen whales (for both oil and whalebone). After the Civil War, the decline in demand for whale oil and the retreating fashion for hoop skirts could have further accelerated the demise of the whaling industry were it not for the latest must-have item of ladies' fashion—the corset. Although corsets had been around for decades, the falling away of hoop skirts promoted a new aesthetic in which women were encouraged to show a long, "hourglass" figure. Thin, pliable strips of vertical whalebone were perfectly suited to the task. The corset fad was born, built upon the feeding mechanisms of countless whales.

The uptick in the importance of whalebone came, in the words of one contemporaneous historian of whaling, as "a gleam of light from out of the darkness of the time. It would seem as though it pointed with no uncertain significance to the fact that this product of the whale fishery was to assume greater importance in the market."[5] Among the whaling fleet the shift in focus from sperm whales to baleen whales also meant a shift in voyage routes and destinations. It additionally required learning the migration patterns of entirely different species of whales. And that meant chasing baleen whales to where they most commonly congregated—the plankton- and krill-rich waters of the Arctic Ocean in the summer. The shift added new layers of complexity and danger to an industry that was already heavily burdened by both.

Now 40 years old, Captain James Dowden of the *Progress* had worked his way up the whalemen ranks in the midst of this transformation. In that time, he would have witnessed the incremental depletion of the whaling fleet at the hands of Arctic ice: The *America* in 1851; The *Marcus* in 1853; The *Mount Vernon* in 1856; The *Newton* in 1857; The *George and Mary* in 1860; The *Henry Kneeland* in 1864; and so on. Just the year before in 1870 the *Japan* and her crew had been presumably lost to ice, although no one was entirely sure. At least three of Dowden's peers had orders to search for survivors during the 1871 season, along with their normal pursuit of whales. In other words, the Arctic ice was a consistent but livable danger with a ship here, a ship there succumbing over the years—enough to serve as a reminder and warning about the volatility of Arctic whaling, but nothing so grim as to discourage it entirely. New London's *Amaret* showed the fleet both the peril and reward

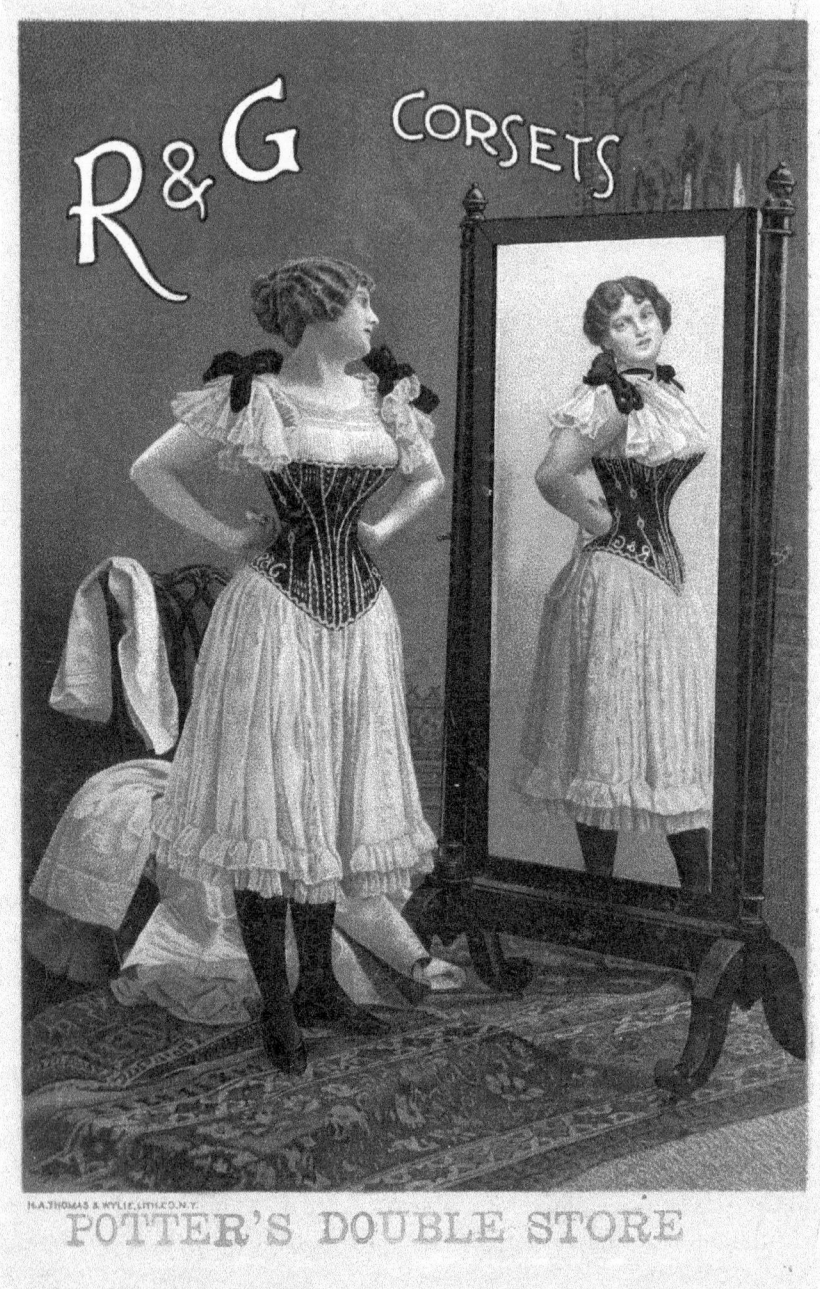

Five. The Arctic, 1871

with this summary of her 1857 voyage: "Frozen into the ice 9 months; took first whale July 1, and by July 22 was full."[6]

The thought of ice and ships colliding might evoke images of iceberg-plagued waters like those that would take down the *Titanic* decades later. In reality, the ice of the Arctic Ocean and Bering Sea was entirely different. "The pack ice is an enormous accumulation of cakes or floes of snow-covered sea frozen ice, of all shapes and sizes" reported William Fish Williams. Twelve years old in 1871, Williams has been born aboard a whaling ship—the *Florida*, in fact—to Eliza Williams. His mother was the very same Eliza Williams who had overcome seasickness to write her descriptive and thorough account of the *Florida*'s four-year whaling voyage. The trip had included the birth of William on January 12, 1859. Eliza and her captain husband continued the tradition of keeping their family together, as both she and her nearly-teenage son again accompanied Captain Thomas Williams, this time aboard the *Monticello*, bound for the Arctic.

Like his mother before him, William Fish William also took a journal and writing utensils with him on the voyage, thus providing a detailed account of his 1871 experiences that complemented his mother's faithful record from the *Florida*'s voyages in the 1850s. William explained that the ice he saw was not towers, but stacks of ice "containing very few whose highest points are more than 10 feet above sea level.... There are no icebergs as there are no glaciers in these northernmost parts of either America or Asia. The pack is not, therefore, in its individual parts imposing, grand or beautiful, but as a whole ... it is a magnificent spectacle...."[7]

To visualize what William was seeing, it is useful to think of the thin ice pack as similar to a 1,000-piece jigsaw puzzle dumped onto a large table. The individual pieces are insignificant and the space between them is easily navigable. But as those individual pieces start to stick together, they become more pronounced and solid. When forced by currents and winds against each other, and as a patch of sea becomes more and more crowded with pieces, light interconnectedness gives way to brute force. Ice starts to slide and pile atop other ice, becoming thicker and denser. Where once the ship might have the upper hand in muscling through a channel lightly strewn with thin ice, thicker ice now becomes the dominant force. Driven by a particularly strong

Opposite: Advertising card for R & G Corsets, undated, including the name of a corset purveyor stamped at the bottom. The "hourglass" shape achieved by the flexible whalebone stays of a corset became the standard of femininity for middle- and upper-class women throughout the late 1800s (Warshaw Collection of Business Americana—Corsets, Archives Center, National Museum of American History, Smithsonian Institution).

The Last Voyage of the Whaling Bark *Progress*

wind or by invisible shifts beneath the water, these denser and denser packs become impenetrable walls at best and battering rams upon a ship's hull at worst.

Still, Captain Dowden and other masters of the whaling fleet in 1871 thought they had a good idea of how the ice behaved, and where they could still find the open or lightly packed channels. A steady stream of successful voyages over the years had produced accumulated knowledge that these men took as inviolable truth—the ice would break apart in the summer as it had previous summers, and the pursuit of bowheads could begin. And so, in January the fleet congregated in the Sandwich Islands (modern day Hawaii) to get outfitted, take on new sailors, and prepare for the 3,000-mile journey north. Forty ships headed north—39 American vessels mostly from New Bedford, and one New Zealand whaler. Almost all of them would never return.

Dowden and the others reached the southernmost part of the Arctic icepack by mid–May. It was both thicker and more southernly than it should have been for that time of year, but the whalers were not overly concerned. They bided their time hunting right whales, waiting for the traditional channels through the ice to clear. They were confined to the dome-shaped body of water that lay south of the Bering Strait, hemmed in by the Chukotka Peninsula on the west and the Seward Peninsula on the east, while St. Lawrence Island ran along the southerly border. The widest distance between the Asian and American spits of land was around 200 miles. For forty whaling ships accustomed to the world as their hunting grounds, this must have felt like being stuck in a bathtub.

It was during these weeks of waiting outside the Bering Strait that three ominous events occurred. First, one of the whalers putting in at Plover Bay, Siberia, on the Asian side of semi-circle made an exciting discovery—the pale face of the *Japan's* Captain Frederick Barker living among the natives. The *Japan's* crew *had* survived her 1870 disaster with the ice, although the ship had not, and now the haggard, bedraggled sons of New Bedford were scattered among various Chukchi villages up the coast. The 1871 fleet made forays to the communities to retrieve their whaling brothers. Eight of the crew had died. The rest were alive, but malnourished and weak. Though this was certainly a joyful reunion, the men aboard the rescue ships could not help but look directly into the skeletal eyes of their comrades to see nature's brutality and mercilessness. This is what happened to men who challenged the Arctic ice and lost.

Second, as each captain tried to test the edges of what was possible in terms of northern progress, some were more aggressive than others. One of the more aggressive was Captain Hayes, whose bark *Oriole* had managed to

Five. The Arctic, 1871

outrun the *Shenandoah*'s path of destruction just a few years earlier during that reign of terror. Possibly convinced his newer and nimbler vessel could get to the whale-rich waters first, he gambled on pushing ahead—and also lost. Stove in the hull by a submerged floe, the *Oriole* could barely hold together a patch long enough to limp back to Plover Bay. She was declared unsalvageable. Her remaining stores and other parts of the ship were sold to another captain. The *Oriole* was the first casualty of the 1871 Arctic season, and she hadn't even made it past the Bering Strait.

Throughout all of this, whalers and Inupiat Eskimos from the North American/Alaskan side of the Bering Sea engaged in trade and discussions. The rituals were well understood between both parties as the Inupiats paddled out to the ships to barter ivory and furs for tobacco and liquor. This had been true of every season that Yankee whalers had come to the far western edges of Alaska. Except this time the conversations had a different tone. It was the third hint that perhaps this year's excursion would be different and ill-fated. The native men arriving in kayaks brought with them the news that the winter would be early this year. A premature refreezing on top of a slow and stubborn melt-off meant that the whalers would be in grave danger if they continued north. Turn around, they were urged by the native inhabitants. Unsurprisingly, the whaling captains and crews were largely unmoved by the warnings of a people they saw as primitive and superstitious. When at last ice broke apart and channels opened in late June, the Inupiat warnings were easily forgotten. Ships began streaming through the Strait and above the Arctic Circle.

Throughout July the fleet worked its way northward in fits and starts. It was a game of "two steps forward, one step back." Winds and weather constantly mutated the ice. At one moment it was an impenetrable barrier of piled up cakes. In the next moment it was a horizon of open channels that must have seemed like Moses parting the Red Sea. When the ships were prevented from pushing ahead, many whalers used the time to hunt walrus, having discovered in recent years that walrus blubber could reasonably pass in the whale oil market and their ivory tusks could also fetch a profit.

Still, even in the 1870s this was a controversial practice and stands as a perhaps surprising bit of early environmental awareness among American audiences, including some whalers themselves. It was well-established and understood that the walrus was the main food source for the indigenous villagers who ringed the Chukchi Sea. In spite of this knowledge, between 1860 and 1880 whalers took an estimated 200,000 walrus. The slaughter prompted outcries such as this 1879 article: "It is computed that for every hundred walruses killed one Esquimaux family is starved to death. What seems to be wanted is a regulation to stop the wholesale destruction of the walrus...."[8] One man who became an advocate for this cause was the erstwhile and recently-rescued captain of the *Japan*, Frederick Barker. Having lived for nearly a year with indigenous men and women who depended on walrus meat, Barker explained, "Should I ever come to the Arctic Ocean to cruise again, I will never catch another walrus, for these poor people along the coast have nothing else to live upon.... I felt like a guilty culprit while eating their food with them, that I have been taking bread out of their mouths."[9]

As July days ticked off the calendar, conditions eventually improved enough for the ships to abandon walrus hunting and return to their true goal. By early August the fleet had rounded the western edges of Alaska and would begin proceeding north and northeast towards Barrow, and Alaska's northernmost tip. The ice was still temperamental though, and kept the whalers confined to a narrow 20-mile-long strip of open water with the Alaskan shore on one side and a sheet of white to the horizon on the other side. They could hear the spouts of whales tantalizingly close, swimming under the ice floes and coming up for air in the gaps that were useless to a whaling vessel. Fog also bedeviled the fleet, preventing navigation even if the ice happened to open. Some tried to send the smaller whaleboats into the floe, rather than the full ship, with instructions to set up camps and carving stations on the ice itself. A few whales were taken this way. Whale blubber and baleen were hauled back to the ship by one whaleboat for processing or drying, leaving most of the crew exposed to the cold and wind while hunting, hauling, and cutting on the ice. If whaling was generally a brutal, taxing profession, this was even worse.

Still, the whalers continued to assume the ice and weather would follow the same pattern they had followed all the previous years. Although the record is not entirely clear, it is possible that during these nearly two months, further encounters with trading Eskimos in kayaks yielded similar, unsolicited advice to what they had heard further south—that the weather this year would be different and dangerous. The fleet did not listen to this round of advice any more than they had the earlier one. And their Yankee confidence was rewarded on August 25.

The expected change finally occurred, and winds shifted to gale-force gusts from the northeast. It was like a hand brushing a swath of the jigsaw puzzle pieces from the table—suddenly several miles of Arctic ocean opened up and nothing stood between the whaling ships and the pods of bowheads feeding in the nutrient-rich waters. It was still a bathtub to be sure—the push probably opened no more than eight miles of cleared ocean beyond the original strip to which the whalers had been held. But after such a long confinement, the ships and barks eagerly scattered into the breach and began several days of successful whaling. Except there were some ships that did not follow into the newly-opened hunting ground. One these was Captain Dowden's *Progress*.

A cluster of ships remained further south and west of the rest of the fleet, including the *Progress, Midas, Arctic, Chance* (the British whaler out of New Zealand and only non–American registered vessel in the 1871 fleet), *Daniel Webster, Lagoda*, and *Europa*. This cluster had begun hanging back even before the conditions shifted on August 25. As the logbook of the *Lagoda* reveals on August 15: "Saw 16 ships most of them working to the North," while the opening of the ice brought sight of another "three vessels under all sail working to the North."[10] In the space of a couple weeks the ice had shifted enough to give the fleet space to spread out; however the *Lagoda, Progress*, and the other five ships did not stream North with the others. In William William's account this seems like a fairly organic, natural distribution: "The fleet was divided into four parts," ranging from a northernmost cluster of four ships to a heaviest concentration in the middle to a few in mostly clearer waters to the south.[11] But was there more going on?

In later years, stories about Captain Dowden at this precise moment would hint at motivations beyond a natural spacing of ships among the available navigable water. "The experience of Capt. Dowden made him apprehensive from certain atmospheric indications of an ice jam, and he accordingly set sail" to the south and west, not north, relayed one reporter in 1893. Other accounts suggested that it was his "not liking the looks of the ice" that would guide Dowden's ultimate decision not to push into the northern breach.[12] One newspaper account in New Bedford herself went so far as to say that the other ships "followed Capt. Dowden's example" after Dowden had intuited danger by "not liking the way the vessel acted, by her feeling."[13] These accounts sat alongside all the other retellings of the coming days in which readers were typically reminded: "When the fleet first arrived in the strait the natives told the masters that they would fare badly.... No notice was taken of this friendly warning, of course."[14] But was that entirely true? Did Dowden take notice of the friendly warnings?

Dowden would spend much of his life after 1871 cultivating the idea that he was both admired by and respectful of non–Anglo-Saxon cultures. An account of his life by the *Boston Globe* in 1911 reported, "He was a guest at the court of King Kalakaua at the Sandwich Islands and speaks the Hawaiian language, as well as being well versed in Esquimau and several of the languages of the South Sea Islands." His obituary in 1915 went even further: "He spent much time in the Sandwich Islands, where he was known in the native tongue as 'The Big Flying Fish.' He was a frequent guest at the court of King Kalakaua, speaking the Hawaiian language fluently. He was also well versed in some of the Eskimo and South Sea Island dialects."[15]

Could it be that Dowden steered the *Progress* away from the northern destinations that others pursued precisely because he actually listened to and heeded the warnings of the indigenous populations? Did being "well versed" in the native languages of the Chukchi rim cause Dowden to recognize something in the dire predictions that others missed? Such questions go unanswered in the official records—nothing in the *Lagoda*'s logbook suggests that Captain Swift was similarly moved or discussed the prognostications with Captain Dowden. In fact, in the *Lagoda*'s case, the crew was busy trying out and collecting the whalebone from an August 24 catch. Perhaps if her luck had been different, she too would have sailed northward on August 25. Still, it is an intriguing possibility that Dowden himself did nothing to dissuade—that his kinship among indigenous tribes gave James Dowden an advantage in interpreting, deciphering, and ultimately heeding the Eskimo advice that "the wind, which was sure to come, would drive the ice against the vessels and wreck them."[16]

With the rest of the fleet out of sight miles to the north and east, the small collection of whalers further south enjoyed a short window of good whaling. Bowheads were caught and processed over the end of August and the first few days of September. Occasionally the *Progress*, *Lagoda* and the rest had to play a bit of cat-and-mouse with the ice, but nothing that impinged on the task at hand. As the *Lagoda* noted, not only was her own crew engaged in taking and harvesting whales, but "several ships" in the cluster were seen boiling blubber in their tryworks.[17] A later report of these days for the southern coterie suggested that "they had just commenced to take whales, which were plenty and available to capture…."[18]

But by the second week of September this Southerly Seven had been forced into tighter proximity by the ice which had begun a steady march back towards shore despite previous years' history suggesting it should be moving in the opposite direction. The corralling by the ice had not been gentle, with especially the *Chance* taking a brutal beating that had turned her into a rather leaky mess. The *Lagoda* was the northernmost of the seven, noting on both September 9 and September 10 that the other six could be spotted to the south.[19] This is why at 2:00 p.m. on September 11 the *Lagoda* was the first to receive a shocking and unexpected visitor from the north. A whaling boat—not a ship, but a six-man whaling boat—came into view. Closer still, it became clear that the boat had a ship's captain at her helm. Captain David Frazier from the now-San Francisco–based bark *Florida* (the very same *Flor-*

Five. The Arctic, 1871

ida that had once been home to Captain Thomas, Eliza and William Williams) came alongside the *Lagoda* and brought with him a terrifying tale.

Captain Frazier had been dispatched from the north two days earlier with a trio of whaleboats. While the seven whalers in the south had certainly endured some weather-related hardships, the situation 70-some miles to the north was far more dramatic and far more dangerous. The retreat of ice on August 25 had almost immediately been followed by its ferocious return. Some of the outmost ships never had a chance to attempt escape and were crushed "like an eggshell" in the first days of September, the *Roman* and *Comet* among them. Those that were able to move back towards shore found no reward for their efforts. Ships fell to the ice like ducks in a shooting gallery: *Awashonks, Eugenia, John Wells*. By September 8, gale-force winds not only drove the existing ice against the fleet like a fist but also added young ice and snow to the building layers. It was obvious no vessel was escaping—winter had come early and had settled in for the season, just as the Eskimos had predicted.

There was a whisper of a channel left by September 9, leading to the south. The captains of the 32 trapped whaling vessels met to discuss options, whereupon it was decided that Captain Frazier would take three whaleboats through the remaining pathway to look for the whaleships and captains that had stayed in the more southernly waters. Fresh in everyone's mind were the haunted, hollowed eyes of the *Japan's* crew who had met a similar fate the year before. Except this time, it wasn't a single ship's crew. It was more than 1,200 souls including a sprinkling of women and children who would never survive eleven months of unrelenting winter until a new batch of whalers hopefully returned the following year. As Frazier began his journey, the fleet also tried one other "hail Mary" idea, hoping that the smallest and lightest of the 32 ships might also navigate the ribbon of semi-clear water down which Frazier took his boats. They failed. Now all hope rested on Frazier.

When Frazier met up with the *Lagoda* at 2:00 p.m. on September 11 he relayed all these facts. He was also making a remarkable admission—that none of the captains believed the fleet could be saved, and that they were all willing to abandon millions of dollars of cargo, ships, and equipment to the Arctic. As the *Lagoda* recorded that day, "he had not any hopes of getting his ship out this season."[20] Even more, he was asking the seven remaining ships to do the same—to abandon the 1871 season and its promise of profits to save those currently marooned to the north. Much would be made of this request in the years to come. A congressional committee noted the "unselfish humanity of the officers and crews of these American whalers in foregoing the objects and profits of their whaling voyage that they might rescue more than a thousand souls from the perils of cold, huger and death...." Of course, the whole reason a congressional committee was commenting at all was the assumption that these captains required remuneration for their selflessness: "the masters of the ships thus addressed determined to abandon their voyages and receive the shipwrecked crews, trusting in the justice and generosity of their government to properly compensate them for their losses and expenses...."[21]

A sort of game of telephone then ensued among the seven intact vessels. The *Arctic* didn't learn of the situation until the 12th, when Captain Dowden signaled to both her captain and the captain of the *Midas* to come over to the *Progress*. Presumably Dowden had been close enough to the *Lagoda* the day before to be part of the original discussions and deliver his famous line to Captain Frazier. David Frazier, the erstwhile captain of an ice-locked whaler, was now working his way back north with news that at least Dowden would wait for them as long as he had "an anchor left or a spar to carry a sail." Fortunately for the fleet, the other six captains had similar, if less colorful sentiments, and the cohort gathered and anchored off the tip of Icy Cape, south of the more incongruously named Blossom Shoals. The waiting began.

When Captain Frazier returned to the ill-fated fleet it was clear they were out of options. It was also clear they were running out of time. Although Frazier's round-trip journey took only from September 9 to September 12, the return was substantially worse than the outbound trip. Yes, he had returned with good news that rescue awaited, but he also had to emphasize that the journey was getting increasingly perilous. Fortunately, the remaining captains and crew had put the time to effective use, sending provisions down the escape route via whaleboats, and discussing the necessary actions to execute a final evacuation. Ever mindful of their legal and financial obligations to the ownership groups back home, on September 12 (after Frazier's return) the captains drafted and signed an extraordinary letter conceding defeat to the elements:

> Know all men by these presents that we, the undersigned, masters of whale-ships now laying at Point Blecher, after holding a meeting concerning our dreadful situation, have all come to the conclusion that our ships cannot be got out this year, and there being no harbor that we can get our vessels into, and not having provisions enough to feed our crews to exceed three months, and being in a barren country where there is neither food nor fuel to be obtained, we feel ourselves under the painful necessity of abandoning our vessels, and trying to work our way south with our boats, and, if possible, get on board of ships that are south of the ice.[22]

The captains also took the time to document their increasingly difficult odds of making the proposed escape with any chance of survival. They first acknowledged that there was "sufficient water for our boats to pass through," but then ensured that future readers understood that these channels were

Five. The Arctic, 1871

"liable at any moment to be frozen over during the twenty-four hours, which would cut off our retreat." One senses that if they could have broken formally and simply said "we are out of time and options," they would have done so. Still, the very real message resonates through the proper, customary language of nineteenth century gentlemen.

Although Frazier had already secured a promise of rescue, they were also acutely aware of the legal and economic bind into which they were putting their southern-stationed peers. So even amid evacuation, they wrote and signed a second formal letter, this one addressed to their rescuing captains:

> ... to make known to you our deplorable situation and ask your assistance.... We realize your peculiar situation as to duty, and the bright prospects you have for a good catch in oil and bone before the season expires, and now call on you, in the voice of humanity, to abandon your whaling, sacrifice your personal interests as well as that of your owners, and put yourselves in condition to receive on board ourselves and crews for transit to some civilized port, feeling assured that our Government, so jealous of its philanthropy, will make ample compensation for all your losses.[23]

On September 14 the procession of whaleboats began, some so loaded down with provisions and humans that they sank to near water level. Even under such dire conditions a certain amount of ceremony remained. Ships raised their American flags before leaving but inverted them as a sign of distress—upside down flashes of red and blue appearing amid the all-too-dominate landscape of white.

The journey was merciless, as if winter wanted to remind all 1,200-plus fleeing souls why they had lost this battle. The wind whipped wickedly, threatening to swamp the boats. The ice that Frazier had feared was building behind him had already coalesced in spots. Whalers often had to break through the young ice or carry boats over it. Some boats—particularly the ones with women and children—chose to create overnight camps on the ice floe the night of the 14th while others pushed ahead through the darkness.

Meanwhile, to the south off Icy Cape, the newly-designated rescue ships waited, and waiting had not been easy. Whaleboats loaded down with provisions had already arrived on the 13th and again on the 15th, along with word that the captains of 32 ships had officially "come to the conclusion to abandon them."[24] The human cargo began arriving on September 16. Unfortunately, the bad weather seemed to have followed the fleeing refugees, creating rescue conditions between whaleboats and ships that were fraught with wind, rain, and fog. The *Lagoda* began bringing aboard passengers by 10:00 a.m. Young William Williams, along with his mother Eliza and father Captain Thomas Williams, were steered to Dowden's *Progress* and brought safely onboard, although as the weather worsened the *Midas*, *Arctic*, and *Progress* all had to

The Last Voyage of the Whaling Bark *Progress*

cut their anchors loose at some point in the day. Dowden's declaration that he would stay "as long as I have an anchor left," was put to a literal test. The *Lagoda* also had to briefly retreat from the scene to avoid encroaching ice before returning and resuming rescue operations. As the nearly 200 whaleboats of the flotilla were emptied of passengers and supplies, each was cast adrift to quickly succumb to the battering floes. The entire enterprise was completed by 4:00 p.m. Dowden's *Progress* took on 188 passengers. The *Lagoda* housed 170, the *Arctic* 176. In all 1,219 men, women, and children had taken the eighty- to ninety-mile journey from the abandoned ships to one of the remaining whaleships off Icy Port. And miraculously, all 1,219 of them survived—there was not a single fatality during the entire exodus.

The surviving ships headed back to Plover Bay, Siberia, where the season had begun. It must have seemed like a lifetime ago. Here they took on water, but the tiny Chukchi outpost hardly counted as the "some civilized port" requested in the captains' letter to their rescuers, nor was it anywhere large enough to accommodate the disgorging of over 1,000 men, women, and children. So from there, the overpacked barks and ships set sail for Honolulu.

"Abandonment of the whalers in the Arctic Ocean Sept. 1871, Pl 4," 1872. Lithograph from painting by Benjamin Russell. The *Progress* is one the largest ships in the image, set in the foreground on the right. Although such details are indistinguishable in this image, those being rescued by the *Progress* would have included the Williams family (courtesy Nantucket Historical Association, gift of Mrs. George W. Allen, 1896.234.5).

Five. The Arctic, 1871

They started arriving in late October, with the poor, beaten-up *Chance* arriving last on November 22.

As soon as the first anchors dropped in Honolulu, the story of the disaster began its rapid spread across America's daily newspapers. The *New York Times* compared the impact on New Bedford to the Great Fire's impact on Chicago.[25] The business community began to quickly assess the impact on oil markets. But the human drama of the events—and the "miracle" that no one had died— became the most oft-told centerpiece of the story, which also proved to have tremendous staying power in the public imagination. The *Boston Globe* noted on the thirtieth anniversary in 1911 that "the story of how the crews were all rescued partly dragging their small boats over the ice and partly sailing them has been told over and over again [emphasis added]." Still that did not stop the *Globe* from telling the story itself one more time in 1915 under the headline: "Death of Capt. Dowden of New Bedford Recalls Arctic Whaling Disaster of 1871," just as the *New York Times* did with an anniversary headline: "Hemmed in by Arctic Ice: The Story of Great Disaster Twenty Years Ago."[26]

The story was also given visual form by New Bedford artist Benjamin Russell and lithographer J.H. Bufford in Boston, who created a popular five-part series of prints called *Abandonment of the Whalers in the Arctic Ocean Sept. 1871*. The illustrations were originally done as part of the owners' insurance claims. Later they were also submitted to Congress in an effort to receive compensation for lost earnings by the six American ships. In the immediate aftermath of the rescue, the whaleship owners had received swift compensation from the federal government of $35 per man saved; however, it would not be until 1891 that they would receive additional compensation for their presumed lost profits for having given up whaling for the season. (The $60,000 awarded by Congress that year was also curiously divided among only five of the American ships, with the *Arctic* conspicuously left out.)

Although the prints were originally conceived as insurance documents, a popular market for them quickly became apparent and the series was sold through newspaper advertisements at least through 1874.[27] *Whalemen's Shipping List*, perhaps somewhat hyperbolically, predicted as early as April 1872 that they "will be sought for by everybody—not only as a work of art, but as a valuable record of the greatest disaster that ever befel [sic] our whaleships."[28]

In addition to the various journalistic accounts and the set of lithographs, there were other widespread versions of events. William F. William's account of his experiences as a child aboard the *Monticello* was first delivered in 1902 and went through several reprintings in popular newspapers and periodicals, helping to keep the tales of Arctic disaster alive into the twentieth century.

The Last Voyage of the Whaling Bark *Progress*

But as the *Boston Globe* headline reveals, it was Dowden who became the most famous and central part of the tale. Certainly his rousing quote of "I will wait for them as long as I have an anchor left or a spar to carry a sail"—whether truly delivered as suggested or not—helped feed the narrative and cement his reputation. However, it is worth stepping back to consider the entire story objectively in an effort to discern if Dowden really was the best figure upon which to focus. He was not alone among the captains who saved the distressed whalers, nor was his ship ever in *unique* danger compared to the others. All the whaleships and their crews risked life and property in executing the rescue. The *Progress* did bring aboard a large number of the refugees, including women and children, but so did the *Lagoda*.

In fact, one might consider Captain Frazier more of the hero and leading man for this piece of historical memory. After all, it was Frazier who set out in a whaleboat into the great unknown to the south, seeking rescue from his fellow captains. Frazier volunteered for that duty unsure if he would make it out, much less find any ships that had somehow remained free of the ice. It was Frazier who convinced those other captains to abandon the whaling season and instead turn their whalers into a flotilla of rescue vessels. It was Frazier who then sped back to the 1,200-plus refugees in time to corral them down a lengthy ribbon of treacherous water (that he had already traversed twice) to the ships he had secured before the ice closed in. Yet scouring the records, Frazier is little more than a footnote, especially in the years after the disaster. It is Dowden who became "one of the most famous of New Bedford whalemen," as he was described in his *Boston Globe* obituary.

But Frazier was also one of the captains who lost his ship and needed rescuing in the first place. And whispers about the captains who abandoned the fleet never really disappeared. Could they have waited it out? Did they give up too soon or too easily? Could they have avoided entrapment in the first place? Speculation swirled, even as whaling agents set about tallying up the insurance claims. And thus it fell to one of the rescuers to be the central hero of this tale, and, in the years to come, the 5'8", 40-year-old New Bedford captain named Dowden would be the man who filled that role.

Six

New Bedford, 1880

Catherine "Kate" Brennan had her hands full, in every sense of the phrase. The 32-year-old Irish immigrant oversaw an extended household crammed into a tiny wooden flat. The minuscule rooms had to accommodate her, her husband, and the six adult boarders who shared their space. The tenement which served as home had recently been built by the Wamsutta Mills Company for employees like her husband George Brennan and the boarders, who all worked at the textile mill churning out high-end cotton goods. Similar long, wooden one-and-a-half and two-and-a-half story double tenements were sprouting like mushrooms around Wamsutta's main mill buildings. They were a sign of the mill's prosperity and growth since its founding in 1848. It was that growth that served as a clarion call to immigrants throughout the 1870s and 1880s, especially Irish, Scottish, and English immigrants in the first years of those decades. In 1880 the Brennans rented at the corner of Austin and Purchase streets in the center of Wamsutta's building boom. A busy hive of hard work and hard conditions, the surrounding area likely felt like a world unto itself. In fact, it was a world that touched nearly aspect of life in the city and spoke volumes about both its past and future. Whether they realized it or not, Kate and George had landed in the middle of New Bedford's biggest revolution.[1]

Not that Kate had much of a chance to stop and consider the changes afoot around her. As the overseer of the three rooms (kitchen, bedroom, living room) that the Brennans rented from Wamsutta, she would have had to coordinate and facilitate the schedules of her six boarders—sleeping, eating, cleaning. The ballet of balancing all the flat's inhabitants' comings and goings rested on a 10-hour work day at the mill. The shifts might have been set even longer were it not for the concession that the Massachusetts legislature had granted to workers only a few years earlier in 1874. The roughly divided day would allow about half of the flat's occupants to sleep and eat during one shift as the other worked, and then vice-versa, with Kate overseeing it all. There was no indoor plumbing. By taking in boarders Kate and George were still responsible for their meals, so Kate would have had to keep pots of stew, cabbage, pickled vegetables, and other inexpensive items at the ready for each shift's return. A little over a decade later—after the installation of indoor plumbing, no less—one commentator called the Wamsutta corporate tenements a "pestiferous excrescence." Still, Kate was hardly alone in juggling the responsibilities of limited resources in squalid conditions. Just a few blocks to the north in New Bedford's poor Ward One, eight people per dwelling was the average.[2]

Kate was also hardly alone in other ways. In the decade prior to 1880, New Bedford's population had grown by 26 percent. Almost all the influx came from immigrants like her and George. The two had come to America in 1874 (for Kate) and 1876 (for George), and been drawn to New Bedford by 1880 for the work to be found in textile mills like Wamsutta in the north part of town, or the Potomska Mills which had started operation in 1872 and lay to the south of town. In the houses along the Brennans' section of Purchase Street the residents' names sounded like the roll call of an Ancient Order of Hibernians' fraternal meeting: Glennon, Sullivan, MacGonigal, Foley.[3]

Her six boarders were also Irish immigrants, except Lizzie, who was born in Nova Scotia to Irish immigrant parents. Assuming Kate's boarding agreements matched the conventions of the period, Sunday would have been her one day off, a time for her, her husband, and the rest of her extended household to walk about a mile south to St. Lawrence Martyr, the imposing Irish-Catholic Church of grey stone that had hosted its first mass on Christmas Day 1870. In the decade since, its growth and prosperity were a further testament to the wave of Irish immigrants coming to the city. Already the parish was talking about opening a school, which would serve the parochial needs of the growing community.[4]

Kate wasn't alone in other ways too. Although she and George had no children of their own, most of her household brood were female. Her female boarders—Betsy (21), Lizzie (21), Maggie (30), and Annie (21)—perhaps also represented something of an extended family. All worked in the cotton mill that practically sat across the street from their tenement, where they could clearly see the clock tower that had been hurrying workers to their shifts since 1868. Then there were Peter and John, both 40 years old. They also boarded in the tenement, and perhaps were the peers and friends of her husband George. One can easily imagine that Kate felt both a mixture of sisterly connection and maternal protection towards the females under her roof.[5]

The vision of young women flooding to New England weaving and cloth mills had been part of the national conversation since before the Civil War. Tens of thousands of young single women had struck out for factory towns—a first wave arriving from America's own farms and country villages, a second wave through immigration. Mill towns had proliferated across the landscapes of the Northeast for at least two generations now, and yet New Bedford was a latecomer to all of this. Although Wamsutta had also opened in the antebellum era, it stood alone in New Bedford for decades and only after the Civil War did its rapid expansion and capitalization begin in earnest. And while Betsy, Lizzie, Annie, and Maggie were just four among hundreds of women who worked in the factory, New Bedford was also unusual in that it still relied on a predominantly male workforce for its cotton manufacturing. The "fine and heavy" cotton yarns that Wamsutta advertised were the result of mule spinning, a physically-demanding production technique using heavy machinery. The process had originated in England to generate fine cloth. The ring spinning machines used in more famous manufacturing centers like Lowell, Massachusetts, could typically be tended by women and children, in part because they had no moving carriage. By contrast, the mule spinners of New Bedford were still mostly men, at least in the beginning.[6]

Unbeknownst to Kate and the people under her charge, all of this was about to change as the 1880s unspooled like one of Wamsutta's thousands of spindles. Among the most significant of those changes was how native New Bedforders themselves felt about cotton manufacturing and the work being done by George, Lizzie, Annie, Peter and the rest of the Wamsutta tenement residents. When founder Joseph Grinnell began raising capital for the Wamsutta Mill in the 1840s, he was met with decidedly mixed results. "The general sentiments of the citizens were in opposition to the introduction of manufactures," remembered one prominent New Bedforder. *New Bedford's Semi-Centennial Souvenir* tiptoed around the issue in typical booster fashion, simply stating that "the task of getting the people of New Bedford interested in the enterprise was a difficult one…." Other sources were more blunt, suggesting that New Bedford had an abundance of capital but much of it was "locked up in the hands of men who have made their fortunes in whaling and do not feel disposed to embark in any new enterprise," in the words of one history, while in another "the great whaling industry" was the "stumbling block in the way of the Wamsutta Mills."[7] Other accounts were blunter still: the wealthy whaling families of New Bedford "clung to the oily tradition of the city, and scorned factories."[8]

Only because he was one of the most powerful men both in the city and in the whaling industry was Grinnell able to push through his rather modest vision of a single mill. Even then success came only by cobbling together the small and reluctant investments of many backers, not unlike the process of organizing a whaling voyage ownership group. When taking up the topic of industrialization in New Bedford, Thomas Bennett, Jr., wrote, "The enterprise was looked at by the great majority of the people as an unsuitable one for the locality."[9]

The 1870s and early 1880s marked a turning point to all of this resistance. Wamsutta added a new brick mill in 1875 and another in 1882. Potomska completed capitalization in 1871, began production in 1872, and expanded in 1877. Between 1881 and 1883 a boom erupted, nearly doubling the capacity of the industry. Grinnell Manufacturing Company, Acushnet Mill, The New Bedford Manufacturing Company, and Oneko Mill were all brand-new enterprises that opened their doors to new business, new orders, and perhaps most importantly, to new workers. They also introduced more and more ring spinning machines, mirroring other cotton manufacturing centers and their reliance on cheaper women and children as employees. Finally, they also represented a sea-change in thinking among New Bedford elites. Money that

"Wamsutta Mill, Circa 1850." William Allen Wall, 1850. This bucolic setting in 1850 was a far cry from the booming scene the Brennans knew by 1880. In the thirty years since Wall's painting, the mill complex and surrounding area had grown into a dense hive of workers, endless construction, and rapid industrialization (courtesy New Bedford Whaling Museum).

Six. New Bedford, 1880

once went towards investments in whaling now flowed freely and willingly to manufacturing.

The capitalization of New Bedford's second mill (Potomska) and the Arctic events that decimated so much of the whaling fleet both happened in 1871. As a practical, economic, or business-oriented matter the two were not linked in any way. News of the Arctic disaster did not begin circulating until November of that year. In fact, the *Whalemen's Shipping List* did not print its full account of events until November 14, 1871.[10] By contrast, the selling of shares and investment opportunities for Potomska's original $600,000 capitalization was largely completed by June 6, 1871, and a second capitalization was sold out by late September. In both cases money flowed to Potomska before investors had any knowledge about one of whaling's greatest financial disasters unfolding thousands of miles to the north.[11]

But the symbolic significance of the timing could not be overlooked, and a psychological connection was made in part because the mill, while capitalized in 1871, did not actually begin operations until July 1872.[12] This meant that the tableau of the Arctic disaster unfolded while the mill's new machinery was tested, floors were constructed, and workers' houses were hastily thrown together. And that gave New Bedforders time to both ponder the significance of the financial wreckage caused by the Arctic disaster as well as time to reevaluate some of the economic warnings about New Bedford that had come in recent months.

When Mayor George Richmond delivered his inaugural address in 1870—a year before either event—he noted to his City Council: "While the whaling business is still pursued with an energy worthy of all praise, and is still the leading enterprise of our city, for reasons beyond our control the fleet has been rapidly dwindling, and the profits of the fishery do not, as once, tempt to new investments in this direction." He went on to note, however, that "wealth is finding channels for profitable investment at home" and that the "whistle of the steam engine heard morning, noon, and night tells where the busy hand of labor is at work."[13]

Mayor Richmond was not alone in this assessment about whaling. Also in 1870 the *Republic Standard* laid out a detailed dossier of the conditions that had launched a sustained decline in whaling, "which in times past has been the great source of the growth and prosperity of our city, but which for the past twelve or thirteen years has been undergoing a gradual but steady falling off. It is worth while to enter upon such an investigation and to see if anything can be done towards arresting the process of decay, if not effecting a return to the former thriving condition of this important interest."[14] By December 30, 1871, and the dawn of a new year, New Bedford had lost 22 whalers to Arctic ice and was barreling towards tens of thousands of new spindles and over a thousand new looms in the cavernous Potomska Mill #1. It was as if Richmond's speech the year before had come from an oracle.

Once news of the losses from 1871 fully settled—along with the insurance checks that compensated as least some of the owners for those losses—the flow of money from whaling to cotton mills began in earnest. The families who had once been synonymous with whale oil and whalebone began to apply their names elsewhere—to director boards for mills along with top executive positions, and even to the names of the mills themselves and their products, such as the Grinnell Manufacturing Company. They were also following well-established patterns developed during the whaling boom of keeping New Bedford money in New Bedford, and keeping outside investors at arm's length. This is what makes New Bedford's transformation so unique—it was a transformation from within. It was a shift not just at the city or neighborhood level, but within the intimate spaces of families who rapidly ceased to be "whaling families" and became "cotton families," often in the span of a generation or even less.[15]

All of this has been duly noted as economic fact in the historical record. In the words of one New Bedford history: "As a result of whalers lost during the Civil War and the Arctic disaster of 1871, the reliable 6% return on the textile stocks looked even more favorable and became the basis for the city's economy."[16] American whaling chronicler Eric Jay Dolan offers this eloquent synopsis: "With every revolution of an oil well pump and every mile of gas line laid, whale oil's days as a major illuminant faded further into the past. By the end of the nineteenth century, whale oil still burned primarily in lighthouses and churches, where spermaceti candles could be found adorning the altars."[17] When the brood of Irish immigrants under Kate Brennan's watch headed off to work each day, they were part of a major economic shift in New Bedford—from lighting the world to clothing it; from baleen to bedsheets.

But the changes were more than just economic. They were also physical and psychological in their scope. Both Wamsutta to the north and Potomska to the south were originally built on quiet stretches of farmland and scrub that were sparsely populated by rural families and occasional herds of cows. As the original mills grew, and still more mills were added in the early 1880s, these former bucolic idylls became the new urban centers, pulling like magnets away from the central core of New Bedford's remaining whaling wharves with their increasingly aged ships, barks, and casks of oil. Some estimates suggest as many as 300 multifamily mill tenements—many literally packed to the rafters like Kate Brennan's home—had been built by 1889. This would suggest many thousands of workers just in mill-built housing alone, much less those double-, triple-, and quadrupled-up in rental housing produced by various independent entrepreneurs.[18] Immigrants like the Brennans would have had little to no connection with the whaling industry that had dominated New Bedford less than a generation earlier. Their neighbors the Glennons or Sullivans or Foleys would have had no reason to head to the wharves to see a ship off or welcome its return; no reason to read the papers for news from the Arctic or Honolulu or Zanzibar.

Not everyone was necessarily pleased by this shift in capital from whaling to cotton manufacturing. Writing on "America's Working People" in 1899, Charles Spahr suggested that whaling had established "a strong middle class community. The cotton industry, on the other hand, had built up fortunes for a relatively few successful managers.[19] When historian Walter Sheldon Tower looked back on this period in his 1907 history of the American whaling industry, he wrote: "The rise of the cotton cloth industry was also a potent factor in hastening the decline of whaling ... had not the cotton mills sprung up, it seems safe to say the whaling fleet would have decreased less rapidly even in the face of increasingly adverse conditions. This is especially true of New Bedford [emphasis added]." Although Tower admitted he couldn't quantify the extent to which whaling suffered at the hands of cotton, it was clear to him there was a correlation.[20] The economist and historian Elmo P. Hohman saw something similar when he wrote his history of the whaling industry in 1928. "To many an old whaleman it must have seemed little short of an outrage when New Bedford herself, suckled and grown strong on oil and bone, denied further capital to her battered old whalers and poured her savings into cotton mills.... New capital ignored the demands and needs of whaling, now obviously senescent."[21] In these histories, the hints begin to emerge that the shift in capital from whaling to cotton did not exist in an emotionless vacuum. Something deeper was afoot.

In the days immediately after the Arctic disaster, the *Whalemen's Shipping List* wrote: "Every once in a while we see it intimated ... that the day of the leviathan is over. Now, we take it that our far-seeing merchants know exactly what they are about, and when they see fit to give up the business, will let the outside world know. We still continue to shine, talk as people may."[22] Even by the 1870s the old Quaker-infused metaphors linking whaling and light were hard to shake, with the promise of "continuing to shine" taking on multiple layers of meaning including the lingering self-importance of the industry itself.

And while many of the former owners and agents took down their signs and folded away their signal flags, others within whaling's tight coterie still could not entirely quit the trade. "The Howland Brothers did not know how to change," observes one author about agents George and Matthew Howland.

The Last Voyage of the Whaling Bark *Progress*

"They noted what had happened; they were certainly aware of the general depression in their industry; yet they pressed on as before." Suggests another author of the Wing brothers, who also remained whaling agents after 1871: "Yet they still channeled a good part of their revenue back into whaling pursuits, which remained the cornerstone of their economic well-being." William Wing was "a whaleman at heart."[23] In 1872, looking back on the disastrous year before, the *Whalemen's Shipping List* lionized those who were not willing to give up just yet: "We may hope for better days to those whose courage keeps them in the way of whaling because they believe we shall see a return of prosperity in this branch of creative industry."[24]

This set up a sort of dichotomy between those who "stuck with" whaling and those who did not. In a 1916 speech, a professor at Haverford College (and a New Bedford native) recalled just such a moment from 1885. Sometime that year Professor Francis Gummere spoke to one of whaling's most revered gentleman, William Taber. When he asked Taber about the origins and reasons for the decline of whaling up to that point, he recalled that Taber mischievously answered, "no whales." But Gummere took from that conversation an entirely different lesson than a display of sly humor: "Ladies and gentlemen, my good old kindly friend said 'no whales'; he did not say 'no whalers.' And here lies such point, or such moral.... There were still whalers at that time in both senses of the term, ships and men." Gummere's story was designed to remind listeners that whaling continued through the 1880s and into the 1890s, even as piles of whaling's familial money was shifting to cotton. "No, my kindly friend could not say that there were no whalers left," Gummere reiterated as he closed out his anecdote.[25]

Within the increasingly shrinking and thus increasingly insular world of whaling, questions about the future of the industry became ever more poignant. Throughout the 1880s the *Whaleman's Shipping List* published their annual "Review of the Whale-Fishery" for the previous year. Reading the entries is an exercise in watching the hopes of an industry fade. Writing of the 1880 season: "...although the wave of prosperity that has swept over the United States has not placed whaling interests in a profitable position, we cherish the hope they may yet be benefitted." By 1884: "Another year has passed, and its results, like its predecessors, have been unsatisfactory and discouraging to those who have continued to risk their capital in the whale-fishery." For 1886: "The year just closed has been without features of special note, except the low prices of oil that have prevailed." And by 1889: "The past year has shown no improvement over the proceeding one in catches or results."[26] What had started the decade as a message of optimism had become nearly a funeral dirge by decade's end.

Six. New Bedford, 1880

It is notable too that the 1880s marked a time of severe depression for whaling agent/owner Matthew Howland, one of those who did stay with whaling. "Nothing new or encouraging," he wrote his son Morris in 1880, "we live in hopes the tide has turned and something, or rather 'little suthing' will be coming in soon. Though we can say truly, nothing very flattering has transpired yet."[27] A year later his angst at the state of whaling was even more pronounced: "The weather for the last few days has been rather gloomy which has had a tendency to depress my spirits and make me feel despondent in our present condition. The whaling business seems running out and doomed to destruction so that we have no business and no way of earning a dollar, so that our case seems a hopeless one."[28] But at the same time, financial moves and decisions that would seem to distance Howland from the "doomed" business of whaling likewise stirred complex feelings. When his firm sold the *George and Susan* in 1881 he wrote to Morrie, "Well, she is gone from us after being in the family 70 years. It seems as though we were to part with everything of a material kind that is near and dear to us."[29]

As Matthew Howland's letters reveal, the rapidly changing landscape of whaling created genuine and strongly felt emotions, sometimes even contradictory ones, among the whaling community. It must be noted that like so many others, the Howlands' continuation in whaling did not prevent crossover investments into cotton. His letters to Morris frequently referenced his dabblings in this emerging market: "I have directed 'Burt' to sell some of my stock in Wamsutta when he can obtain 120 or more," he wrote in 1883.[30] But his words about whaling also belie an assumption that Gilded Age capitalism in New Bedford was simply a dispassionate and detached pursuit of returns and profits. Once these very real human emotions are realized and recognized, the shift of money from whaling to cotton raises interesting questions of loyalty and even culpability.

When Zephaniah Pease wrote his history in 1918, he suggested that it was the "whaling industry which betrayed her [New Bedford]," a decidedly strong choice of words that Walter Sheldon Tower and Elmo Hohman—who argued that any betrayal was at the hands of cotton manufacturing, not whaling—would have vehemently denied. "The business community and the capitalists began to turn from the whaling industry," offers another history, the phrase "turn from" suggestive of turning one's back, while Clifford Ashley—one of New Bedford's most revered chroniclers—more forcefully suggested that the city had "relinquished her birthright too easily."[31] Among even this small smattering of histories we can see both fierce emotions and the disagreements that emerge.

The Last Voyage of the Whaling Bark *Progress*

Yet another recounting of New Bedford's past suggested, "Our attractive and romantic past was not only forgotten. It was actually repudiated. We were on with the new love. We were irritated to be reminded of our long continued amours with the old. The Board of Trade actually became incensed because someone alluded to New Bedford, the new born queen of fine textiles, as an old whaling town."[32] From the very beginning of this shift, there were different prisms through which to see both its character and cause. Pease acknowledged as much in another of his histories, this one for the centenary of the Merchants National Bank: "The old was dying, the new was being born—both experiences were painful."[33]

The argument certainly preceded the 1870s and 1880s. As noted before, an earlier generation of whaling's elite had mostly managed to hold off manufacturing; however, that is not to say the decision went uncontested or was a forgone conclusion. As early as 1863 the fiery Rev. William Potter of New Bedford's First Unitarian Church declared from his pulpit that whaling as an industry was in fact holding New Bedford back, not propelling it forward: "The business is a relic of barbarous pursuits which the civilization of the present age is outgrowing. … and it is almost a providential that the business is by and by to cease. Even if it should recover for a moment, no city should be dependent on a single enterprise." The men who sat in the pews of Potter's congregation included William J. Rotch and William W. Crapo, both men who owed their initial fortunes to whaling.[34] It is more than a coincidence that both men would also go on to turn their backs on the "barbarous pursuits" of whaling to become some of New Bedford's most prolific builders and investors in cotton manufacturing. To return to Zephaniah Pease's declaration of a betrayal—who betrayed whom?

Historian Margaret Creighton offers another intriguing interpretation of the shift. A cultural emphasis on the value of cleanliness increased throughout the 1800s, as clean clothes, clean faces, and clean hands came to also represent moral virtue and superiority. For decades New Bedford had ignored the snubs and sneers about filthiness that had been directed at whaleships and whalemen from all corners of the world. "Even though whalemen cleaned themselves and their ships before sailing home," Creighton writes, "merchant seamen and navy crews were witness to their at-sea appearance and carried their unfavorable impressions back to the United States…. One whaleman summed up the looks of whaling work, and by association, the industry's reputation in genteel company: 'Everything,' he said, 'is beshit.'"[35]

While at one time New Bedford had little choice but to turn a blind eye (and nose) to the grime of whaling, now they had a new industry rising in their midst. Even as portrayals of cleanliness gained greater and greater culture cache, New Bedford's genteel society had a second option to present to the outside world—one that didn't reek of blood and guts and gore. And while cotton mills themselves were hardly bastions of pollutant-free cleanliness, the cotton products of New Bedford were much more conducive to a narrative of purity and health. Drawing from their history of putting out "fine" cotton products since the mid–1800s, Wamsutta was able to declare to women in 1921: "Your great-grandmother bought Wamsutta sheets…. Their pure, white crispness endures many launderings."[36]

Of course, not everyone saw things the same way. There was the question about how all this investment in cotton was changing the city. Many argued it was changing it for the worse. The new physical geography of New Bedford—with its recently emerging magnets of cotton to the north and south, and its dying core of whaling in the center—was matched by a massive shift of demographics. Matthew Howland—one of those who stubbornly continued whaling's business—admitted twice to his son in the span of two months, "It is rather lonesome in the office," and again, "I find it rather lonely in the counting room."[37] And while Howland engaged in an increasingly solitary walk from home to wharf, he and his peers still occupied several square blocks of stately, patrician homes within the heart of New Bedford. Though many no longer invested in whaling, the social elite families who had financed the industry remained physically close to the increasingly deserted waterfront streets and wharves. These were still the homes that whaling built, whether or not the occupants continued to join the ownership groups of New Bedford's dwindling list of annual voyages.[38]

By contrast, the north and south were exploding as working-class neighborhoods, the centers of which were often built by the mills themselves. But even the cheap tenement housing that the Brennans knew so well could not be put up fast enough to keep pace with immigration. Soon the Irish families would be joined by larger and larger ranks of French-Canadians and then Portuguese families. Unlike the whalers and sailors who left New Bedford for years at a time, these new arrivals required housing, groceries, and other services year-round. By 1880 the city had already seen 26 percent growth, and even this was a mere shadow of what was to come. In the next two decades the population of New Bedford would double.

Clifford Ashley looked on these changes and turned the "cleanliness" narrative on its head: "For three miles along her harbor front brick and cement factory chimneys belch forth blackened smoke where once tall masts and white sails raked the skyline. This was called progress."[39] Nor was the rift subtle. Outsiders too could see the split among the two communities of

The Last Voyage of the Whaling Bark *Progress*

cotton and whaling, with a visiting *Boston Journal* reporter noting in 1871: "Cotton mills now monopolize the north and south ends of town. Iron and copper works employ large numbers of men. Steam manufactories are attracting flocks of young folks, whose heads are filled with anything but Puritan notions. So the city is in transition and the venerable captains and retired merchants are often surprised with the strange sights around their comfortable mansions."[40]

Throughout much of its history New Bedford had earned a reputation for liberalism and egalitarianism, in part because whaling had gathered a cosmopolitan mix of individuals from around the world and brought them to the city's shores. The Quakers' anti-slavery focus and advocacy for abolition had likewise fostered a desire to live up to the ideals of equal treatment for all. Perhaps because of this, the influx of immigrant workers did not immediately trigger the levels of nativism and suspicion that one could see in other cities in the 1880s. In fact, early strikes by mill workers were often met with support by New Bedford citizens, and New Bedford consistently ranked among the top cities for per capita expenditures towards poor relief prior to the mid–1880s.

But as the numbers swelled and more non–English-speaking immigrants took the place of early English and Irish migrants (like those in Kate Brennan's tenement), it was not just the cotton industry that drew rebukes, it was the workers themselves. Commentators began to see the words "immigrant" and "poor" as largely synonymous, suggesting "the eager search for candidates for naturalization at every election is saddling many needy cases

Cyanotype of Acushnet Mills, 1887. The massive footprint of New Bedford's cotton mills created an entirely new urban landscape for the former whaling city. However, not everyone viewed the changes as "progress" (courtesy New Bedford Whaling Museum).

upon the city."[41] And others linked the immigrants' arrival as a threat to whaling's importance and place within the fabric of New Bedford. Without a hint of irony William Crapo—now wholly a cotton man—himself lamented that the foreign-born "know and care little about our early history" even as "the old stock is seriously diminishing in numbers."[42]

There were other, more subtle ways that New Bedford's elite whaling cohort (or in many cases, former whaling cohort) attempted to make their own imprint on the immigrants who seemed so physically and emotionally removed from whaling. In 1870 Rachel and Matthew Howland erected a non-denominational, public chapel for mill workers, on Purchase Street near the Wamsutta Mill. It opened January 13, 1871, at a cost of $7,000.[43] They named it the Howland Mission Chapel, and Kate Brennan, Lizzie, Annie and the rest must have walked past it every Sunday on the way to the Irish-dominated Catholic church that served their faith and community needs. In fact, the Chapel was so seldom used that in 1884 Rachel Howland began investigating how much she could recoup by liquidating it.[44]

The Howlands' son William opened the doors of the New Bedford Manufacturing Company in 1882 to produce cotton yarn. He also, by 1889, built 50 worker cottages based on the latest theories of industrial paternalism, utopian idealism, and moral uplift for the working class. Some have argued that this was partly inspired by watching his father Matthew Howland outfit his whaling voyages with the best food, gear, and supplies.[45] Somewhat ironically, when New Bedford's remaining whaling interests organized the New Bedford Board of Trade, Matthew Howland, last of a dying breed of whaling agents, wrote to his son Morris that the organizing "ought to have been done 20 years ago" and that "it seems to me they ought to have had some of the Manufacturers [as] members."[46]

And in 1887 New Bedford enacted a law that required minors working in the mills to be able to read and write English, or otherwise be forced to attend evening school. The Wamsutta Mill alone had nearly 200 of these new compulsory students, who once safely ensconced in the classroom not only rehearsed language skills, but likely served as a captive audience for good civics lessons on New Bedford's history of Yankee ingenuity.[47]

In 1888 the *Boston Globe* quipped of New Bedford, "When the whaling industry began to wane some one [sic] prophesied that before many years the grass would be growing in the streets of this old city. But how different is the fact."[48] New Bedford had indeed dodged an economic bullet and created a second act for itself. Still, the world of Kate Brennan and her boarders was a *new* New Bedford, one in which whaling could be forgotten or largely ignored. And the two worlds were moving further and further apart. Fourth of

The Last Voyage of the Whaling Bark *Progress*

July whaleboat races in the harbor were cancelled, replaced with boisterous mill neighborhood celebrations attended by Irish immigrants. When Zephaniah Peace wrote about William Howland's worker homes in 1890 he noted that the mill communities were "practically independent of the old whaling port. These communities have their stores and their amusement halls, and there is more activity on the streets at these extremes of the city than there is elsewhere."[49] Steamships carrying cotton and other cargo took over where whaling ships once anchored along the docks. Purchase Street's north-to-south axis linking the two cotton manufacturing hubs became the city's new commercial thoroughfare.

In time there would be more open discussion about the psychological conflict between whaling and cotton. One lengthy article, entitled *New Bedford: City of Whales*, especially drew upon the fault lines. "'The old city then,' I said in my heart, 'is still the city of the whalers,'" wrote the visitor in the early 1900s. "But when I put the question to the Board of Trade, they answered me with something of pity for my old-fashioned ideas. 'No,' they said, 'whaling is gone by. It does not pay nowadays.'" And yet this same visitor would attend a meeting later in his stay in which a whaling captain's presentation was part of the program: "No wonder the walls of the building seemed to rock in the applause that followed, and to open for a little, and let the choke of the spindle lint out and the strong rich smell of the full sea in." The author's turn of phrase to distinguish between the "choke of the spindle lint" and the "strong rich smell of the full sea" is hardly a subtle one. And still, the dichotomy between whaling and cotton was pushed to even greater relief elsewhere in the piece, layering additional thoughts about the city's masculinity and virility to the argument: "The capital that came rolling in from the southern seas, wrested from the mightiest animal of the world by the keen emprise and audacious daring that molds men, big men, gathers interest from whirling spindles tended by little men with pale faces, steep shoulders, and furtive eyes."[50]

In a few more years a local editorial would question New Bedford's proclivity to promote cotton over whaling: "If whaling were 'soft-pedalled,' and if by some magic, the whole nation could be forced to forget that New Bedford had a whaling past, what would be left of our fame? Would strangers then be more willing to think of us as a great cotton manufacturing community? As a matter of fact, they would hardly think of us at all…."[51]

But all this reflection would become clearer with time and distance. For now, few words were shared that explicitly dealt with a major transformation not only of economy but of identity and a sense of self. Except, there was this: During one particular holiday, a whaling ship was hauled into the Acushnet

Six. New Bedford, 1880

and set on fire as a nighttime spectacle. Though it was described as a "glorious sight," the crowd didn't cheer or whoop, but remained silent, as if watching a funeral pyre.[52] The brain could articulate all the dispassionate and disconnected reasons why cotton mills rose as whaling fell; but what the heart said for those in New Bedford's shrinking whaling community seems to have been another matter entirely.

Seven

Chicago, 1892

Henry Weaver had fair fever. Ever since April 25, 1890, when President Benjamin Harrison signed the bill officially declaring Chicago the site of a world's fair in 1893 (after a bruising battle on Capitol Hill to wrest that crown from New York City), the city of Chicago had been abuzz with both work and anticipation. Chicagoans knew the world was watching. While many cities offered their congratulations and support for Chicago's undertaking, others perched on the sidelines with little attempt to hide their schadenfreude should the Midwest metropolis fail. As one wag in Memphis wrote, "It will be especially interesting to know how the Windy City is to fill the bill, with all her workmen on a strike and a horde of Anarchists eagerly waiting for a chance to raise a rumpus."[1] Chicago's civic leaders and industrial elite knew they had to stage the best and most spectacular of everything in order to silence the skeptics who still thought of their city as a cultural backwater.

No doubt Henry Weaver considered himself one of those "industrial elites." At just 40 years of age, Weaver had already cemented his place in Chicago society as a millionaire coal baron. His was the sort of plucky, up-from-your-bootstraps narrative that Chicagoans loved. The son of a long-established farming family near Niagara Falls, Weaver headed to Chicago after cutting short his time at Yale University due to an unspecified illness. Arriving in the Windy City in 1881, Weaver set to work establishing himself in the coal industry, building up a little company that modestly moved about 10,000 tons of coal annually. More importantly, Weaver knew the importance of networking, hand-shaking, and back-slapping in a town like Chicago. Early on he developed and nurtured relationships with those who ran the city and its municipal infrastructure. By 1892 Weaver, Getz, and Co. was supplying coal to a vast web of Chicago's hospitals, penitentiaries, police stations, asylums, pump works, City Hall, and the Cook County courthouse. The amount of coal Weaver supplied to Chicago alone had grown to more than 600,000 tons per year. Following the example of other industrialists, Weaver also worked towards vertical integration in his supply chain, and secured interests in sev-

Seven. Chicago, 1892

Advertisement for Weaver, Getz, and Co. From *"There She Blows"* (Chicago: Arctic Whaling Exhibit, Co. 1893).

eral Indiana mines that produced his coal, and in the railroads that carried 150–200 carloads of coal each day.[2]

Thus, when Chicago won the rights to host the 1893 fair, Henry Weaver got to work making sure his city impressed the world during its debut on the international stage. He also got to work thinking about how he could profit from the upcoming exposition. The most obvious path was the one that had built his fortune—supplying coal to the city, or in this case, the "White City" as the fair came to be known. Unfortunately for Weaver, the fair organizers had other ideas.

When the call for bids to supply fuel to the fair power plant went out in early 1892, the response came not just from several Chicago-based coal companies, of which Weaver had the lowest bid, but also from Sinclair Oil. In fact, Sinclair had been assuming it would win the contract for a month, something that Henry Weaver openly chaffed at in the Chicago press. "The Standard Oil company has been figuring for this contract for six months," Weaver groused,

The Last Voyage of the Whaling Bark *Progress*

"with full information in regard to the appliances, boiler construction, etc. and the specifications for the boilers were drawn on the basis of oil as a fuel long before the fuel was advertised for." In the end, Weaver's protests were for naught. Sinclair won the contract after, in the words of one trade publication, "some necessary discussion and some slight investigation of the feasibility of installing an elaborate and costly system of coal and ash."[3] Blocked from his more traditional path to profits, Henry Weaver began to consider other ways he could both contribute to and gain income from the Columbian Exposition. The answer came in the form of a whaling bark from New Bedford. It is no small irony that the man whose plans were thwarted by the petroleum oil industry found his way to the city that had once lit the world.

Whaling had enjoyed a place in some of America's smaller exhibitions and fairs prior to the Columbian Exposition thanks mostly to the efforts of the still-young Smithsonian Institution. In 1884 the "National Museum" did double duty, staging a whaling exhibit at the Southern Exhibition in Louisville, Kentucky, from August to October, and then relocating the display to New Orleans in time for the December 16 opening of the World's Industrial and Cotton Centennial Exposition. These exhibits were, in turn, largely recycled from what the museum had sent in 1883 to the Great International Fisheries Exhibition in London.[4] The display consisted of a full-sized whaleboat with an array of harpoons and a whale gun. The whaleboat sat in front of a large painted canvas entitled "Capturing a Whale," which showed a crew in dramatic action. The Smithsonian worked to further educate audiences on the mechanics of whaling through "a model of a whale-ship with a whale alongside, showing the method of stripping the blubber and trying it out on the vessel's deck; and by paintings of whaling scenes."[5]

And as plans unfolded for the Columbian Exposition, whaling once again became the subject of modest attention, despite its continued and precipitous decline. William Nye made sure to secure good space for New Bedford's Nye Lubricants, and his display of whale and fish oils for watches and chronometers promised to be an eye-catcher. By the 1890s, the Smithsonian had moved on in focus to large natural history and ethnographic cases for the fair, but New Bedford stepped in to ensure that a similar display to the Smithsonian's earlier work was staged in the Fisheries Building. Perhaps they could even improve upon what the National Museum had accomplished. The arrangement would include irons, harpoons, knives and samples of whale oil and soap. There were drawings, models of whaling ships, and "groups of sperm-whale jaws, a large walrus head, an assortment of whalebone, and specimens of Arctic animals."[6] There was also a completely equipped whaleboat suspended over the main floor, although at such a height the details of

Seven. Chicago, 1892

the equipment might have been lost. New Bedford's leaders had an enthusiastic and sympathetic partner for their endeavors. The Chief of the Fish and Fisheries Department of the fair, Captain Joseph W. Collins, had spent some time as a young man in Massachusetts. Captain Collins wrote to planners back in New Bedford, "I note with much interest the very complete exhibit you have sent and shall feel much pride and interest in having it properly installed.... The brave old city by the sea has well acquitted herself and it shall be no fault of mine if the fame she has acquired in every land and on every sea does not receive proper notice at this exposition."[7]

Then there was the gentleman in Provincetown—a whaling captain named Amos Chapman—who wanted to capture a live whale and display it at the Columbian Exposition. Chapman's idea generated both excitement and a sizable amount of derision. The *New York Tribune* poked fun at the plan for several paragraphs before ultimately concluding with this advice: "We like the idea and wish them godspeed. But they must remember that it is easier to coax a whale than to drive him.... If the chairman can get up close enough to pat the whale on the head and let him eat out of his hand he will have gone a long way toward solving the problem."[8] This kind of mockery notwithstanding, Captain Chapman's plan suffered not from a lack of support, but from a lack of whales. His window during the summer of 1892 closed without a capture and he returned to port empty-handed. The aquarium space fair organizers had hoped to use for a whale went to sharks instead.[9]

But back in New Bedford, folks had been germinating an idea for some time that was on a much more ambitious scale than just a whaleboat, harpoons, and a few models. They wanted to send a whole whaleship to Chicago. The first to get the idea off the ground in a meaningful way was Captain Henry Clay. In his mid-fifties in 1891, press accounts typically painted Captain Clay as "one of those retired seamen who make so large a part of the population of New Bedford."[10] Indeed, he had been the captain of several voyages before making the shift to owner/agent at a time when many were leaving the field, including David Kempton by 1877. His fleet at various times through the 1870s and 1880s had included the *Golden City*, the *E.B. Conwell*, the *Union*, and starting in 1885, the *Franklin*, a schooner. Now idle among the quiet wharves of New Bedford, it was the *Franklin* that Clay suggested fixing up and sending to Chicago for display. At least one account suggested that her crew could use a "dummy whale" to demonstrate "how the work of cutting in is done."[11]

The *Franklin* seemed like a fine choice. In fact, it had made news just a couple years earlier when on July 15, 1889, it came to the rescue of a Boston Fruit Company steamer that was completely engulfed in flames off Cape

The Last Voyage of the Whaling Bark *Progress*

Hatteras. Arriving in the middle of the night, the *Franklin* was able to save all but two of the crew as well as a female passenger from Scotland. All were in the water due to the steamer's lifeboat overturning on launch. Henry Clay was awarded two-thirds of a $3,500 compensation by Congress for the rescue which caused "the schooner to leave her cruising grounds and break up her voyage."[12] The captain and crew received the other third.

Clay did much of the groundwork to make his idea of a whaleship at the fair a reality. He submitted the paperwork necessary to apply to fair organizers, and subsequently fielded their questions when they expressed doubts that the diminutive schooner was actually a whaler. He initiated the conversation as to whether a fee could be charged once the *Franklin* was on display within the fairgrounds. And Clay mapped out the timing, noting that an early departure not only ensured avoidance of ice but also maximized revenue by allowing intermediate stops for display in cities like Montreal and Buffalo. "It is the true boast of her owner that he has never yet undertook any business affair which did not prove successful," wrote New Bedford's *Republican Standard*. "He seems to have put his heart into this latest scheme, so when the World's Fair is over the stout little whaler may have become famous not only at home but abroad."[13]

By the end of 1891 it seemed that Clay's idea was largely a *fait accompli*. Newspapers from Philadelphia to Chicago, and thanks to syndication, as far away as Bismarck, North Dakota, and Atchison, Kansas, were running articles describing the *Franklin* and its future trip to the world's fair.[14] But then in January 1892 new information began trickling out of New Bedford. "At a meeting of the Board of Trade today," reported the *New York Times*, "on the World's Fair exhibit a proposition was made to fit out the old whaling bark *Progress*...."[15] The whaleship now under discussion was the very same *Progress* that Captain James Dowden had used to rescue nearly two hundred souls in the Arctic ice over twenty years earlier.

Indeed, both the *Progress* and the *Franklin* had been in port for months, as were many other relics of whaling's better days. Yet one can almost imagine the conversation in the closed room of the New Bedford Board of Trade that January. If a schooner, then why not a bark, which is much bigger and more impressive? After all, fair organizers had already expressed doubts that the petite schooner was really a whaleship. *The Daily Mercury* hinted as much, later writing: "it was proposed by some of the more enthusiastic members that a large bark or ship be fitted out to represent the city [emphasis added]."[16] Going even further in the comparison, if the *Franklin* was famous for rescuing a few merchant seaman off the coast of North Carolina, how much better would it be to display the ship that rescued hundreds of whalers in the Arctic?

Seven. Chicago, 1892

One could even add to the discussion that Henry Clay had already received reward money for owning the *Franklin*. Why not let someone else have a turn of good luck?

Or maybe it was the fact that George F. Bartlett, who owned the *Progress* now, was a lover of history and historic preservation. Did his passion for history stir the hearts of the Board of Trade when he articulated his vision for what a whaling bark at the fair could accomplish? Or perhaps the more salient detail was that George Bartlett was one of the founders of the Board of Trade, which just happened to meet in the offices of I. H Bartlett and Sons. As they sat in Bartlett's chairs and looked out his office windows perhaps they could not help but swing their decision behind the *Progress*.[17] Whatever the case, the *Franklin* was out, and the *Progress* was in.

With the matter of the vessel determined, the Board of Trade continued to strategize about their plans and next steps. They developed a series of organizing committees, completed complicated plans for stock capitalization companies, and calculated itemizations of what the enterprise might cost. It was all very thoughtful, thorough, diligent work, but none of it would matter without the most important missing factor: George Bartlett needed to find a buyer.[18] They found one in the Chicago coal baron named Henry Weaver.

In May of 1892 word went out that the *Progress* had been sold. The *Whalemen's Shipping List* barely noted it at all, simply marking the *Progress* in their table of whaleships as "May 4, 92, sold and withdrawn."[19] But rumors had been circulating that a sale was in the works, in part because George Bartlett needed to reassure the buyers that a canal trip to Chicago was truly possible. "If it is found that her draught is not too great, or if her keel can be cut down sufficiently to enable her to pass the locks on the St. Lawrence river, Chicago parties have agreed to purchase her," reported the *Evening Standard* on April 21, 1892.[20] The auguries on the matter apparently boded well. The buyers were in New Bedford to close the deal within two weeks.

Initial details were vague, but early reports suggested that the trip to Chicago would begin soon, that the journey would be via the St. Lawrence River, and that the *Progress* would be put on display as "exhibit of the old-time whaling trade." Some articles detailing the news used phrases like "a syndicate of Chicagoans" or "Chicago parties," but Henry Weaver was clearly the face and the name that both the Chicagoans and the New Bedford Board of Trade put forward when discussing the deal.

It is somewhat curious that a Chicago coal merchant, who was born and raised on a farm in upstate New York, came to find out about—much less care about—purchasing a relic of New Bedford's whaling history. In those months between January and May, it seems likely that the New Bedford Board of

The Last Voyage of the Whaling Bark *Progress*

Trade would have used their network of business associates and contacts in Chicago to secure a buyer (or buyers) for the *Progress*. It is possible that they even hired a dedicated agent to travel to the Windy City on the project's behalf. Whatever the origination point might have been, it is clear that the road to success led through the dark paneled and plushly carpeted rooms of the Union League Club of Chicago.

Founded in 1871, the Union League Club was "ground zero" for the entire Columbian Exposition enterprise. It was no secret that the Club had been instrumental in securing Chicago as the host city, wielding its members' considerable influence to sway that decision away from New York City. Daniel Burnham, the famed designer and Director of Works of the Columbian Exposition, was a member of the Union League Club, as was the Director-General George R. Davis, the Exposition President Harlow N. Higinbotham, and Higinbotham's first and second vice-presidents, Ferdinand Peck and Robert Waller.[21] As it happened, the Union League Club of Chicago also had a member who was becoming increasingly active and prominent in the Club: Henry Weaver. By 1893—the year of the Columbian Exposition—Weaver would be elected to a three-year stint as a Club director, along with seats on the House Committee and the Reception Committee. He had only just joined the club back in 1886, and that was a mere five years after Weaver had arrived in the city.[22]

New Bedford may have been surprised at Weaver's interest; however, if Weaver himself was not an obvious choice, he *had* married into a family of history enthusiasts that surely had some influence on his frame of mind. Henry's wife Addie Guthrie Weaver was a true lover of history. In 1898 she published *The Story of our Flag*, a patriotic children's book filled with stories about George Washington and Betsy Ross, along songs and poems such as "A Patriotic Song Dedicated to the Sons, Daughters, and Children of the American Revolution." In her preface, Addie writes: "For some years the Author has been interested in the history of our First Flag and its fair maker Betsy Ross, and fortunately, through a family relationship with one of the descendants, became familiar with much of the family history." She dedicated the book to "Mr. George Canby of Philadelphia, and Mrs. Sophia Campion Guthrie of Washington D.C., grandson and great granddaughter, respectively, of Betsy Ross."[23]

And then there was Ossian Guthrie, Addie's uncle, perhaps the "family relationship" that introduced her to her familial history.[24] One of the more substantive articles about the *Progress* in May 1892, immediately after the sale, was published by Chicago's *Daily Inter Ocean*. In it, Ossian Guthrie, who had just returned from New Bedford, served as a bit of an advance man for the

Seven. Chicago, 1892

Chicagoans. "I am not at liberty to say who the interested parties are, or how much the vessel cost. Some of the gentlemen went East themselves, and when they return in a few days they will tell you all about it." Ossian Guthrie was an interesting choice to be part of Henry Weaver's entourage in this endeavor, and not only because he was Weaver's uncle by marriage. Guthrie had served as chief engineer at the Bridgeport Pumping Works, one of the largest water control works at the time and the main feeder of water into the Illinois and Michigan Canal. He was even more famous for his advocation of a drainage canal to drain off sewage from the Chicago River. More importantly as a linkage to Weaver's interest in American history, Ossian Guthrie was also an amateur historian, eventually becoming an honorary member of the Chicago Historical Society. He wrote a book about his father, the inventor of chloroform, documenting the road to that scientific discovery. Guthrie also extensively studied and wrote about the explorer Père Marquette.

From this perspective it is perhaps not so strange that the *Progress* caught Weaver's interest. We may never know if his first introduction to the idea was from talking about fair-related investment opportunities among his friends or from reviewing an actual prospectus that perhaps circulated through the rooms of the Union League Club in the early months of 1892. However, we do know that it also caught the attention of another of Weaver's fellow club members, William D. Preston. Preston was even more involved in the Club than Weaver, serving as Treasurer in 1892 and as a Club executive alongside the high-ranking Columbian Exposition officer Ferdinand Peck. At the time he was popular enough among his Club peers to appear on both the 1892 "Regular" and "Opposition" tickets as the Club's candidate for Treasurer.[25] Preston worked as one of the top executives of Chicago's Metropolitan National Bank, whose President, another Club member, also served on the fair's Board of Directors. Preston himself was deemed important enough (or wealthy enough to afford the placement fee) to be included in the "Representative Men and Women of Chicago" section of the *Columbian Exposition and World's Fair Illustrated*.[26] Naturally, Weaver's coal company kept its bank accounts with Preston's Metropolitan National Bank.[27]

Henry Weaver and William Preston were united in another way—both had grown up along the eventual route of the *Progress* to Chicago. Weaver was born and raised near Niagara Falls in upstate New York, while Preston hailed from Detroit, his home until college. Perhaps as they considered a prospectus and plans for buying the New Bedford whaleship, they also envisioned the impact it would make on their boyhood homes as it stopped along the way.

There was one additional early investor in the *Progress*, another Chicagoan and fellow member of the Union League Club named Edward H. Turner.

The Last Voyage of the Whaling Bark *Progress*

Turner had a lower profile than his two compatriots. Though a member of the Club, he never held office; nor was he involved in any of the fair committees. He was also likely less wealthy than his co-investors. He co-owned a wholesale dry goods firm with his brother, which was rated as worth between $45,000 to $75,000 in 1890—a far cry from Weaver's coal company and mines, or Preston's bank. Still, Turner was hardly a pauper among princes, owning a Greenwood Avenue mansion with seven bedrooms, five fireplaces, and a large library.[28]

It was these three gentlemen—Weaver, Preston, and Turner—who would file the incorporation papers for the Arctic Whaling Exhibit Company with the state of Illinois on June 14, 1892. The purpose of the company was "the exhibition of apparatus articles and processes illustrating the whaling industry and incidental to whale fishing and whale oil production." The next day, the Illinois Secretary of State empowered the three Chicagoans and Union League Club members to serve as commissioners of the corporation and to open the books for $50,000 worth of capital stock to be subscribed to through five-hundred $100 shares.[29]

Despite their enthusiasm for the project, the fact remained that they were the outsiders to the world of whaling, reliant on New Bedford herself not only for the whaleship, but everything that was needed to turn it into an attraction at the fair. Thought had already been put into this. Back when news of Henry Clay's *Franklin* was circulating in late 1891 the articles typically mentioned the "museum of objects in the whaling line, from a whale's head and bones to harpoon and knives."[30] Of course, the Clay and his schooner ultimately lost out on the chance to take a whaler to the fair, but most of Clay's original ideas remained intact, including the idea of collecting a museum of whaling objects for display on the ship. This fact was likely little comfort to the man who had been muscled aside from his original idea. The same day news of the *Progress* sale was given front-page prominence in New Bedford, Henry Clay announced that his *Franklin* would be broken up for whatever "old iron and other junk could be obtained from her hulk."[31]

With the *Franklin* out of the way and the ink drying on the contract detailing Bartlett's sale of the *Progress* to Weaver, New Bedford's elite got to work in pulling together *their* idea of a whaling museum. The Board of Trade created a fund for object purchases, intending to collect "every variety of implement used in chasing the ocean monsters, such as harpoons, lances, harpoon guns, spades for cutting out, tackle, trying-out furnaces, and so on."[32] Another account suggested: "She would be supplied with try-pots set, all whaling gear, and four boats, and a captain and at least a crew of six, all of them practical whale men."[33] When Ossian Guthrie spoke to reporters back

Seven. Chicago, 1892

in Chicago, he too relayed that "She will be fully equipped with bomb guns, harpoons, and all the other implements used in whaling—in fact, she will be completely rigged out ready to start on a whaling expedition."[34] The *New Bedford Evening Standard* went even further. Visitors to floating museum would first be given "a short description of the whaling business" and then would be treated to a fake sighting of a whale from the masthead, the lowering of boats, and the capturing of an imaginary whale. "It is proposed to have practical whalers at hand to show to visitors every appliance connected with the business and explain its use."[35]

To those who lived in New Bedford, these were obvious and reflexive choices. Of course a whaling museum should have "every variety of instrument" and "every appliance connected with the business." The more, the better! After all, hadn't the Smithsonian Institution done largely the same thing when they presented displays at those earlier expositions and fairs including London, Louisville, and New Orleans? But take a step back and the idiosyncratic nature of these assumptions become a little easier to spot. Is this the kind of display that would really be a popular ticket seller back in Chicago? The Smithsonian may have approached their exhibit with a similar eye for encyclopedic thoroughness, but the Smithsonian had not viewed their whaling exhibit as a money-making enterprise.

These early ideas for the museum reveal a certain tunnel vision endemic in New Bedford. No matter that whaling was an extremely complex process with a series of discrete and interconnected stages. No matter that executing that process required specialized knowledge and a unique vocabulary at each step along the way. No matter that the mention of bomb guns and harpoon guns suggested that whaling had evolved with modern times in ways that stripped away the drama of countless whaling adventure stories. No matter that "every appliance connected with the business"—knives, harpoons, lances, spades—that would be so proudly displayed and explained were implements of a bloody, gory, filthy process that most Americans knew little about.

Still, those involved gave themselves little time to reflect on their enterprise. Ossian Guthrie suggested that the *Progress* would sail less than two weeks after her purchase. It was an extremely optimistic prediction, and, ultimately, unrealistic. The local paper put the needed turnaround time at closer to a month. But the sense of urgency sent everyone into a frenzied pace: "A large gang of riggers and sail makers were doing their best to get the *Progress* ready for sea, between decks and in the cabin; carpenters, painters, and other mechanics were working as fast as they could."[36]

Amid the flurry, hints could be seen that a second narrative was already starting to emerge. Not only would the *Progress* be an authentic representation of whaling, but it would be a reminder about one of whaling's most enduring stories and dramas—the Arctic Disaster of 1871. In fact, when Henry Weaver and his syndicate created their company to purchase and exhibit the whaleship, they dubbed the enterprise the "Arctic Whaling Exhibit Company." Although the incorporation documents made no mention of the "Arctic" portion of the name, press accounts did remind readers about the *Progress*'s remarkable rescue. "The most stirring event in its history was the part the vessel played in rescuing ship-wrecked crews from the Arctic regions," reported the *Chicago Daily Tribune* in their story about the sale, although they erroneously suggested that said events unfolded "among the icebergs of Greenland."[37] The *Boston Globe* fared better in the geographic accuracy of their reporting, pointing out: "She is the only one left of a fleet of five that in 1871 brought home 1200 seamen from the 37 whalers that were crushed by ice in the Arctic Ocean."[38]

The focus on the events of 1871 also meant that objects gathered for the floating museum needed to help tell that specific story. "Among the many articles that will comprise this exhibition will be whale bones taken by a captain who was lost in the terrible disaster of 1871," reported the *Evening Standard* (although it is a curious claim since no captains were lost), would be "polar bears skins and a valuable set of pictures taken by a man who was in one of the ill-fated ships and was saved and carried 90 miles across the ice to the vessel in open water with his pictures intact." Beyond the physical objects, early reports also held out a tantalizing possibility. "An old whaler, who sailed her on many of her voyages, will come with her," suggested Ossian Guthrie, while other accounts teased, "Probably the crew, who will explain things to visitors, will be ancient mariners some of whom years ago sailed out of New Bedford on the *Progress*." Clearly those involved were hoping to capitalize not only on material objects related to the rescue, but potentially on some of the veterans from 1871 and other voyages as well.

These two narratives—a didactic recreation of whaling with an emphasis on authenticity, and a theatrical recounting of the dramatic events of 1871—were not necessarily incompatible goals for a whaling museum. But it should also be recognized that they were definitely not the same thing, and they had the potential to pull the entire purpose of the museum (today we might even call it the "mission statement") in different or competing directions. One relied on the accuracy of interpretation and demonstration through objects; the other on the narratives of storytellers told upon the stage of the *Progress* herself. One fit the classic mold of museums as filled with cases and artifacts; the other hewed closer to the "living history" museums of today. Intermingled with this interplay of possibilities were Henry Weaver's own vision and

Seven. Chicago, 1892

plans for the *Progress*. The timing is significant because these ideas and potential avenues were starting to become co-mingled with something else in Henry Weaver's life that was entirely removed from questions of whaling. Weaver was increasingly involved in plans for one of the most extravagant, spectacular, and massive theatrical productions ever conceived. It was called the Spectatorium.

By the time Weaver, Preston, and Turner officially incorporated under the Arctic Whaling Exhibit Company name in June 1892, another world's fair-related company had also been incorporated, one called the Columbian Celebration Company. This too was a company organized to create a world's fair attraction, but on a much larger scale than the *Progress*—a 10,000 seat theater. Dubbed the Spectatorium, the venue promised to produce "the largest-scale pictorial illusions ever witnessed on stage" and it would all be created by one of the most famous stage producers in America, Steele MacKaye.[39]

Henry Weaver would also become heavily involved with the Columbian

"The MacKaye Spectatorium" (artist rendering). Childe Hassam, 1893. Steele MacKaye's ideas for the Spectatorium would have made it the most ambitious theatrical venue in America. The fact that Henry Weaver was an early investor and booster of the enterprise suggests a bit about his interest in entertaining spectacles compared to didactic museum displays (MS Thr 412 [Box 49, Folder: Correspondence–Spectatorium painting], Houghton Library, Harvard University).

The Last Voyage of the Whaling Bark *Progress*

Celebration Company—joining as a company director and offering his endorsement in August after "a dinner at the Union League Club." But even before that, his two enterprises were overlapping. When the Arctic Whaling Exhibit Company filed a list of shareholders in July it included Albert Tucker, who built Steele MacKaye's models for the Spectatorium and Egbert Gillett (manufacturer of Magic Yeast), who was the first Chairman of the Columbian Celebration Company's board.[40]

What attracted these men as investors was that MacKaye was not interested in producing just any run-of-the-mill spectacle. He wanted to recreate the entire pageant of Christopher Columbus discovering America, including a voyage across a man-made interior "ocean." According to one scholar, MacKaye desired to create a "temple of entertainment" in which he could "present historical stories that would enthrall" his audiences. As MacKaye spun his visions of "supreme struggles," "exalted aspirations," and "heroic endeavors" to investors like Henry Weaver, how could Weaver not be thinking about his other investment—another maritime-themed attraction, no less— and the level of drama *it* would contain?[41]

In the dichotomy between these two visions of what the *Progress* would or could become, one can sense an already bubbling tension. New Bedforders were envisioning an exhibit that would speak to the business and mechanics of whaling. To them the romance and appeal of whaling was inherent and presumed; that the thrill and lure of adventure would be manifest in the harpoons and lances and imaginary whale chases. They could turn this project over to a Midwestern coal baron with no whaling experience in part because they assumed the display itself would carry the day.

This, of course, was the assumption of those who had lived in "The Whaling City" their entire lives, and who had grown up for generations with an evangelical zeal for bringing light to the world. This was the assumption of those who had largely chosen to remain in New Bedford's central core of patrician stone houses, within sight of the wharves, counting houses, and businesses built on whaling. Now, as whaling was dying, the enterprise of memorializing and preserving whaling—both the actual industry, as well as its mythic place in the American psyche—took on greater importance. Particularly for those families that had turned their back on whaling for the more reliable profits of cotton goods, this was a chance to ensure that the origins of their fortunes were honored. The local paper in writing about the *Progress* articulated just such a sentiment for all of New Bedford to read: "all who take an interest in our present city, which is an outgrowth of that industry, will feel a spirit of exultance that its old fine business is to be so worthily represented."[42]

Yet the *Progress* and the "spirit of exultance" that it presumably produced

Seven. Chicago, 1892

for whaling was both an opportunity and a curse. Coincidentally, Herman Melville had died only a few months prior to the sale and outfitting of the New Bedford whaleship. His ill-received *Moby Dick* was just beginning to recirculate among a few East Coast literati in the early 1890s. But the reasons for its initial failure should have served as a warning to those scrambling to prepare the *Progress* for display—it was seen as "an ill-compounded mixture of romance and matter-of-fact."[43] Even as men in New Bedford tried to remind Americans that "old whalers are full of romance and valuable history from stem to stern," there was perhaps a blind spot among the Board of Trade and others—a blind spot created by decades of romanticizing and deifying, whaling as the guiding light of industry, home life, and America itself.[44]

Adding in the delays, it still only took 35 days after the sale of the *Progress* for her to be ready for departure—on June 8, 1892. The plans, ideas, and concepts that had gone into sending a whaling ship to the world's fair were about to give way to the reality of audiences, spectators, and viewers' own decisions about what a museum to whaling should be.

Eight

Lake Ontario, 1892

Captain Gifford had a theft problem. Not from his crew, a diverse mix of men whose backgrounds spanned oceans and continents. Gifford was, after all, still captain of his ship and discipline was both expected and received. No, the problem was tourists. Tourists were stealing the objects that had been hastily gathered in New Bedford and from the surrounding communities that dotted the Massachusetts coast. Although the carpenters in New Bedford had done a fine job converting the lower decks of the *Progress* into a series of galleries for museum displays, they had neglected one detail—lockable cabinets. "When the whaling schooner *Progress* reached Oswego," reported the local newspaper, erroneously changing the whaleship from a bark to a schooner, "it was found necessary to provide show cases to hold her relics.... So great was the desire of the thousands of sightseers, who were permitted to go on board, to secure a memento, from the large collection that many were stolen."[1] Captain Daniel W. Gifford, now 53 years old, had survived whales' death flurries, sharks, ice, tropical diseases, and gales, but now he was bedeviled by souvenir-seekers.

Gifford had first been made a captain because of the Civil War's brutal, final blow to the New Bedford whaling fleet, and David Kempton's small collection of whaleships especially. The time since that frantic promotion had been eventful. After sailing Kempton's *Spartan* out into the Atlantic as her captain, Gifford found himself ravaged by fever throughout the two-and half-year voyage. He wasn't alone. A high number of crew—nine of his men plus himself at one point—had been fighting "fever and ague," an old-fashioned name for malaria, since the whaler's layover off the Brazilian coast.[2] Eventually the decision was made to return to New Bedford with a sick captain. "Thus the Spartan commences her road towards home," noted the logbook on June 1, 1868. "God grant that she may safely reach there in due season and that we may all once more behold our dear native land."[3]

Gifford returned to the rank of first mate for the next few years, serving in that capacity aboard the *Kathleen* for her 1880–1884 voyage. Apparently in these intervening years he had also picked up a nickname of sorts:

Eight. Lake Ontario, 1892

"Mr. Gifford has a hard name, and is known throughout the fleet as 'Bloody Dan,' but I can find no fault with him," recounted one his crewmates from that voyage in a 1936 whaling memoire entitled *Harpooner*. Later the crewmate modestly adjusted his memory in this rather pulpy but nicely illustrated volume, writing: "This rough officer, known throughout the whaling fleet as 'Bloody Dan' Gifford, for the first six months out from home never had a civil word for me. He used to call me a packet rat and worse, but for the last three years he has made amends and favored me many times."[4] The author also noted Gifford's propensity for sickness: "Whenever we go in to the coast, he always gets the fever and has to take quinine all the time. Out at sea he is a strong, healthy man." He also recorded Bloody Dan's own words on the matter: "I'll be mighty glad to get into cool weather again for I always feel better then. When a fellow goes on the coast of Africa in those low lying lands, he's likely to get the fever and anybody that gets it once will be likely to get it again."[5] It seems certain that Gifford was describing frequent relapses of malaria.

When the *Kathleen* returned, Gifford was elevated to captain for her next two voyages, 1884–1887 and 1887–1890. He was 45 when he took command for the first time since the *Spartan*. Sixteen years had passed, but Dan was a captain once again. When he departed in 1884 he had only been home with his wife Lucy for four months. When he departed again in 1887, his time with Lucy was even shorter. Over the course of ten years he saw his wife, and his only child Lester, for a total of seven months.

He also spent the early years of the 1890s in San Francisco, which had become something of a "New Bedford West" for the remaining whaling fleet, as the whale fishery's center of gravity moved from the Atlantic to the Pacific. For a time, Honolulu reigned as the Pacific center of whaling operations, but after 1880 the hub shifted again to San Francisco. Steam whalers were introduced into the American fleet in 1880, and the adoption of steam in San Francisco was one reason it became a growing whaling port when other places were giving up the pursuit. It is from San Francisco that the Wing brothers sent the *Charles W. Morgan* (today the only surviving American whaleship in the world) out on short nine- and ten-month voyages to the Pacific. Although he had been given the taste of captainship on the *Kathleen*, Gifford joined the ship's 1891 season as first mate.

It was through these voyages that Gifford remained in the orbit of the Wing brothers for over a decade. J. & W.R. Wing was by now one of the oldest ship-owning firms in the whaling business. Like David Kempton, the Wings had entered the whaling business at its peak, but unlike Kempton, the Wings seemed unable to give up on the trade. The *Kathleen* was one of only 19 ships

The Last Voyage of the Whaling Bark *Progress*

to leave port in 1884 and, like so many others by the 1880s, sailed out under the J. & W.R. Wing flag. By one analysis, "from 1880 until 1910, two years after William Wing's death, the Wing fleet was the largest single fleet of whalers in the world."[6] Their firm was doing more than anyone else to extend the longevity of the industry, and aboard the *Kathleen* and *Charles W. Morgan*, Gifford would become part of that narrative.

Which is perhaps why he was selected by the New Bedford Board of Trade to captain the *Progress* to Chicago. Born in 1839, Dan Gifford had spent a life in whaling, starting at sixteen years of age.[7] The trajectory of his own adulthood mirrored the rise and fall of the industry itself. Now in his early fifties, Gifford, like the whaleships back in New Bedford, looked his age. His formerly lean face, robust mustache and beard, and piercing eyes had given way to clean-shaven but droopy jowls, thinner hair, and a decidedly wider girth. But Captain Gifford was still a whaling captain when fewer and fewer men could claim that title. Now New Bedford entrusted Gifford and a small crew to see the *Progress* safely to Chicago.

Still, the roles of captain and crew aboard this particular voyage were not at all what Gifford was accustomed to, and their work was unlike any voyage he had captained before. For example, the *Progress* did not hoist her sails once she left New Bedford on June 8 and entered the Atlantic Ocean. Instead, the whaleship was towed up the east coast by the coal-powered tugboat *Right Arm*, a decision likely made to preserve the sails' pristine whiteness and new condition. It turned out to be a prudent choice given the tug and whaleship encountered a "short northeaster off Caple Sable which … gave the *Progress* some lively exercise." The journey to Quebec City was accomplished in seven days, with the *Progress* arriving on June 15, 1892.[8]

Slipping past the citadel and the recently broken ground and construction site of the Canadian Pacific's Chateau Frontenac, the *Progress* continued to glide down the St. Lawrence River into Montreal the morning of Saturday, June 18, taking up residence at the opening of the Lachine Canal. His work complete for the time being, Gifford turned things over to Fitch Crane, the man who would put into motion the publicity machine that would accompany the *Progress* as she made her way to Chicago.

Ezra Fitch Crane was a strange addition to the mix of millionaires and social elites who became involved with the *Progress*. Part of the "syndicate of Chicagoans" that Ossian Guthrie had described back in May 1892, Crane styled himself as Henry Weaver's agent and perhaps even protégé. It was Crane who was selected from the Chicago group to remain behind and journey with his wife aboard the *Progress* to Chicago. Crane was also heavily involved in the logistical matter of gathering the objects which formed the

Eight. Lake Ontario, 1892

The *Right Arm* towing the *Progress* to Quebec City, 1892. From the very beginning, the *Progress*'s trip to Chicago would be unlike any whaling voyage. That included being towed up the Atlantic coast rather than using her sails. This was done to preserve their pristine condition (copyright Mystic Seaport Museum, 1973.899.171).

museum, although many others would be credited in this task as well, suggesting it was something of a group effort.[9]

When in the presence of millionaire Henry Weaver, Crane was barely mentioned, relegated to the shadows and subsumed under the coal baron's starring role as the owner of the whaleship and leader of the Arctic Whaling Exhibit enterprise. However, away from Weaver, Crane often asserted himself more forcefully. Described as a "young businessman," Fitch Crane would sometimes claim that the *Progress* had been his idea, as he suggested to a reporter in Montreal when the *Progress* arrived that Saturday morning in June. "The gentleman who is responsible for this bit of enterprise is Mr. E.F. Crane of Chicago. He got the idea that a whaler in the upper lakes would prove interesting and profitable and imported the scheme to Mr. Henry E. Weaver, of the same city, and they both, putting their heads and pockets together, secured the prize which is assured to prove a success to them as she did to her former owners."[10]

It all sounded very good on paper; Crane was presented as an equal partner and social peer of Henry Weaver. Unfortunately, what Crane had shared with the Montreal reporter was almost entirely a fabrication. One major problem with the claims in Montreal was that Crane didn't "get the idea" of a whaler at the fair. It was a proposal that originated with Henry Clay in New Bedford nearly a year and half earlier. For a *Harper's Weekly* article later that summer, Crane softened his story slightly, suggesting that he was merely the first to *hear* "of the offer to sell the *Progress*, and fit her out for a whaling voyage to Chicago." Again, that article claims that he brought the idea to Henry Weaver, a story also told in New Bedford the day before the *Progress*'s launch.[11] Thus if Crane had a tendency to self-aggrandize and stretch the truth, Montreal was not his first opportunity to do so. Searching the newspapers of the era, he is even credited with stopping a forest fire while driving around Cape Cod in the couple weeks prior to sale of the *Progress*.[12] It is easy to sense that Fitch Crane was as adept at promoting himself as he was at promoting the *Progress*.

E. F. Crane certainly was a partial owner of the *Progress*, albeit with a modest two shares of the original 500 offered. But his social class and standing was far different from Weaver's. Like Weaver, Crane hailed from upstate New York. Fitch's father was a horseman, and because Fitch was an only child, all his father's knowledge as a horse driver and trainer fell to him. He would remain connected to horses for much of his life, later dabbling in horseracing and auctioneering. In 1890 Fitch signed on to be postmaster of the tiny village of Cascade, New York, a post he relinquished in less than year.[13]

Instead, he headed to Chicago where his knowledge of horses would land him a job with Leroy Payne's Palmer House Livery. Described as "an equine palace," Leroy Payne's livery was a far cry from generic city or back alley stables. By the 1890s Leroy Payne himself was rich enough to invest in multiple properties in downtown Chicago along with a hotel in St. Louis. Crane joined Leroy Payne's Palmer House Livery staff and worked his way up to the legitimately significant role of superintendent of the livery's operations.

In the day-to-day execution of his duties, Fitch Crane was given access to some of Chicago's richest inhabitants and visitors at the exclusive Palmer House Hotel.[14] Strikingly handsome, Crane no doubt made the most of his charm and his position to introduce himself to Chicago's elite.[15] This was likely his conduit to Henry Weaver, who eventually became his business partner in the *Progress*. However, this was vastly different from being a social peer

Eight. Lake Ontario, 1892

or equal to Weaver, no matter how much Fitch tried to spin it otherwise to the press. Fitch Crane and his wife Charlotte could be assigned to spend weeks on the *Progress* as it was moved from New Bedford to Chicago, precisely because of his differing social strata from the rest of the *Progress*'s investors.

What was clear—no matter his pedigree—was that Fitch Crane was ambitious and blessed with the gift of gab. It made perfect sense that Crane was the one who took over from Captain Gifford once the *Progress* was safely ensconced in Montreal. It was Crane who "took the *Witness* [the *Montreal Daily Witness*, the city's major English-language newspaper] representative in hand" for a tour of the *Progress* and ensured a lengthy column came out the evening of the whaleship's arrival. It was Crane who promised that sightseers would be given an opportunity to see the bark before her departure. And it was Crane who began to craft the messages about the *Progress*. In doing so, he was the first to give voice to exactly what the floating museum was meant to be.

In this, Fitch Crane largely stuck to the original vision of the New Bedford sellers who wished to memorialize their industry. Crane's major "talking points" included the fact that the *Progress* was a "lucky ship," i.e., a good producer of whale oil and whalebone, a claim backed up by a detailed accounting of how much of each product the *Progress* had produced in her lifetime. The role of the *Progress* in the 1871 Arctic Disaster story was, of course, repeated, as was mention of Captain James Dowden. But perhaps most tellingly, the *Witness* reporter was given an unvarnished presentation of whaling and the tools of the trade: "The vessel is a regular museum," the newspaper's readers would learn that evening, "and contains a large collection of whaling curiosities. Whaling guns, old and new, bone lances and harpoons are to be seen there in great variety. Odd and cumbersome they are, indeed, but they could not be made otherwise, and prove so effective as they are in securing the sea monsters." The article even included illustrations and detailed accounts of four whaling implements.[16]

This was the first test of New Bedford's vision for a whaling museum, and it seemed to pass with flying colors. The reporter had been shown harpoons, lances, and whaling guns and came away proclaiming that the *Progress* was "a venerable volume of history and romance, in which each timber is a chapter, each rope a page."[17] It was exactly as the New Bedford Board of Trade—indeed the entire whaling community, past and present—would have hoped for, i.e., that the romance of whaling would be self-evident and innate.

But even within this first test of the overall purpose and goals of the New Bedford-conceived whaling museum there were hints that not everyone was experiencing the same narrative. The *Montreal Herald*, another major

The Last Voyage of the Whaling Bark *Progress*

English-language newspaper in the city, barely acknowledged the *Progress* beyond a pair of short paragraphs over two days. In one it reported the ship was attracting a good amount of attention; however, it suggested in the other that "a glance at her time-worn bulwarks will carry the mind back in a time when whaling was a far more dangerous employment than it is now."[18] The takeaway that whaling should be easier today than in decades past wasn't exactly wrong—the introduction of harpoon guns and bomb lances had given whalers some modern advantages—but it was a strange dichotomy to draw. It also hinted at future questions for the floating museum. Was it a faithful representation of whaling in the present day or the more dangerous whaling of old before the advent of modern killing methods? In museum parlance—what era was the whaleship interpreting? The article was proof that even at her debut, the *Progress* could generate reactions that deviated from the original script of an uplifting display to whaling's greatness.

Meanwhile, major work was being undertaken, which surely did not help Fitch Crane's campaign to awe viewers with the *Progress*'s mere presence. New Bedford's contribution to the world stage was at this moment literally being stripped naked.

In 1892 there was only one route that allowed ship access from the Atlantic Ocean to the Chicago World's Fair, and it passed directly through the Lachine Canal. While the canal allowed ships to avoid the unnavigable Lachine rapids, it was hardly a perfect match for a heavy, high-masted bark like the *Progress*, which also had to pass under the bridges that spanned the canal. So, the whaleship was shortened, and lightened, and then lightened some more, and then lightened yet again. Her topmasts, sails and rigging were all removed, as were the whaleboats. The ship's ballast soon followed, along with the exhibit materials. And finally, the keel—the underwater backbone of a timber running from stem to stern—was removed in drydock.

With his whaleship sufficiently denuded, Captain Gifford once again found himself reliant on being tugged, this time with all of his gear and accouterments in a shallow barge tied to the stern of his ship. "It was slow sailing," suggested one account in a bit of understatement considering the voyage took a day longer than anticipated. "[T]hey passed the little town of Lachine, through the beautiful Lake St. Louis, then by Cornwall, with its hive of mills, and finally through the famous thousand islands with its wealth of scenery and multitude of cottages and palaces."[19] Other accounts put less emphasis on the scenery and more emphasis on the roughness of the journey for the *Progress*. As one crew member recounted decades later, "it was just barely possible to get the old bark through in the shallow spots, and even with her shallow draft she dragged in many spots through the canals."[20] Scrapes and

Eight. Lake Ontario, 1892

all, on June 25, 1892, they arrived at their next destination of Ogdensburg, New York—out of the canal and into Lake Ontario.

Here Fitch Crane again swung into action. Newspapers in the county dutifully reported that E.F. Crane had arrived with the *Progress* in Ogdensburg, and that he "has charge of the enterprise," a moderate stretch of the truth given that Captain Gifford was, well, captain; however, it seems safe to assume Gifford had no desire to spend his time dealing with reporters,

"Passing through the St. Lawrence Canals." G.A. Coffin. The *Progress* is nearly unrecognizable in this illustration of its passage into the Great Lakes. Nearly everything that could weigh her down during passage through the canal system has been shifted to the flat bottom barge behind her. From *"There She Blows"* (Chicago: Arctic Whaling Exhibit, Co., 1893).

The Last Voyage of the Whaling Bark *Progress*

so Crane's version likely passed unnoticed. Reports offered details similar to Montreal's, including the recitation of the whaleship's productivity numbers, the label of a "lucky ship," and accounts of events in 1871. The process of whaling remained front and center in the article printed upon the *Progress*'s arrival, just as had happened in Montreal, noting that "all the whaling gear of both the past and present" was on display and the exhibits offered "quite a museum of articles used and obtained by whalemen."[21] Crane also had the benefit of some advance publicity, with the *Daily Journal* also carrying the story of the *Progress*'s departure from New Bedford a few weeks earlier: "She is a typical old New Bedford 'spouter,' probably the last of the old time fleet, and fitted as she will be with boats, try works, and the rest of her whaling gear will be an interesting exhibit."[22]

But then something happened that Crane clearly did not control. Ogdensburg's local newspaper ran a second article two days later. "The old whaler *Progress* was visited yesterday by several hundred people," it noted. "She is an odd looking craft compared with the usual lake vessel … she is short and looks dumpy." The author noted that she had gone through a stripping to get through the canal and "it is said that when fully sparred her looks are much improved."[23] Still, this was hardly the image Crane wished to project, and it was a reminder of how visuals and the superficial markers of whaling were core to the entire enterprise. In order to succeed, the *Progress* had to look the part. It was one of the very reasons Henry Clay's schooner *Franklin* had been shunted aside—it didn't look like how people thought a whaler should look. Ossian Guthrie had assured Chicagoans of the *Progress*'s visual pedigree back in May: "The vessel is the best whaling vessel I ever saw. It is not a tub and not a clipper, but a genuine merchant whaler that has seen service."[24] But in Ogdensburg, that narrative stumbled precisely because the *Progress* had come out of the canal looking rather tub-like.

The June 27 article would go on to highlight another problem. "A good deal of amusement was caused by the comments of visitors on the craft and her equipment. One said that he always supposed the oil was extracted from the whale by squeezing the fish through rollers and he was looking around anxiously for the machine." Here, laid bare in only its second stop, was one of biggest problems with a whaling museum—whaling was a complicated, multi-step, confusing process riddled with inscrutable tools and impenetrable jargon. And while the crew looked the part, right down to their ability to tell yarns and "their arms tatood [sic] in the manner laid down in the story books," those story book versions of whaling did little to prepare the public for how complicated the process truly was.

This impromptu second article out of Ogdensburg is a terrific little gem

Eight. Lake Ontario, 1892

for articulating early on the challenges facing the *Progress*. So much relied on the visual cues of whaling drawn from a generation's worth of dime novels, serialized stories, and morality tales like *Peter the Whaler*—the very "story books" the writer mentions. As the *Progress* sat anchored in Ogdensburg, some things matched the engrained tropes, such as tattooed sailors spinning yarns, while others fell short, like the lack of topmasts, sails, and rigging. But most importantly, it was already being revealed that those stories and illustrations had left the public completely unprepared for understanding or appreciating the realities of the inner-workings of an industry.

The man who assumed his whale oil came from rolling a whale through a press might have amused the crew and even the reporter, but how many readers of the newspaper that evening were still left wondering, "So how do you get whale oil from a whale"? And more importantly, did they care enough to learn the answers from a floating museum dedicated to the whaling industry? One ominous sign of the answer to the latter question came in another local newspaper article a few days later. Describing a replica of Christopher Columbus's *Santa Maria* that would be coming through the same canal route shortly, the author suggested: "Residents along the banks of the St. Lawrence are likely to be afforded an opportunity of seeing another vessel which in point of interest will knock the old whaler Progress out of sight."[25]

As the *Progress* was put back together—hoisting in ballast, getting her rigging back aboard, restoring her fittings—another problem emerged: theft. As the museum cargo was reloaded from the canal barge back onto the ship, it became obvious that items had gone missing. Gifford deduced "that while in port at Montreal a number of their most valuable small relics were stolen." And so, after the *Progress* left Ogdensburg on June 30, it made an additional stop down the coast of Lake Ontario to Oswego, where "special cases" were made, presumably with strong locks. He may not have been able to sail his ship like he wanted, but Captain "Bloody Dan" Gifford could make sure his cargo was protected. Cases bought, the *Progress* charted a course (this time under the power of the tugboat *Haskell*) to Buffalo.[26]

The *Progress* arrived into Buffalo on July 2, 1892, having successfully been towed around Niagara Falls through the Welland Canal. She would stay on display through the Fourth of July holiday, although weather conspired against bringing out large crowds for the first couple days. She "will be an object of interest," suggested the *Buffalo Express* to those looking for activities during the Independence Day holiday, "to all who are inclined to be curious, as she is a full bark-rigged three master and has a complete whaling outfit on board."[27]

A few days later the newspaper ran a Sunday feature on the whaleship,

The Last Voyage of the Whaling Bark *Progress*

giving the *Progress* front page prominence, including a photograph of the ship at the Buffalo Breakwater. The extended attention in the Sunday edition had some interesting points of focus. One topic that clearly resonated with the reporter was the challenge of getting the *Progress* as far as Buffalo. In describing Captain Gifford and the crew it was noted that they were a "picturesque crew of salts, most of whom apparently had a poor opinion of fresh-water sailing." The article went on to detail the logistical trials and tribulations: "The Progress had a hard time in getting through the St. Lawrence and Welland canals. She was stripped of everything that would lighten her, even her spars and some 12 inches of her keel being taken off to give her less draught. Her ballast was shipped up by rail." Now that she had finally cleared all the canals (and their overarching bridges), it was possible to restore the topmasts to their full height and glory. At least the *Progress* was once again able to look the part she was supposed to play, as was featured in the large front-page photo.[28]

Similarly, the author circled back to how the crew definitely looked the part. In addition to being described as "picturesque," the crew "includes several powerful fellows who have braved the dangers of a whaler's life and is of the usual cosmopolitan character, a hardy lot of men who have grown brown and grizzled in the service, and who now delight to spin yarns of exciting days when they were younger." The article also kept a strong focus on the whaling industry and the implements of whaling aboard: "There is a rack of harpoons and lances that have seen service half a century ago, and their edges are as keen as though the captain expected to run into a school of spouters on the trip up the lake."[29] If Fitch Crane was keeping score, the Buffalo feature surely outweighed the unfortunate piece from Ogdensburg.

The next tow—this time by the *City of Berlin*—was to Detroit, a run made in about 25 hours. The stay was only a few days, but was surprisingly contentious. "Since she left the Atlantic she has been visited by thousands of persons" Detroit's *Evening News* groused, "but on arrival here Capt. Gifford appears to have become soured on strangers, as he positively declines to permit anyone to board her." The man nicknamed Bloody Dan had apparently reached his limit. There would be no more visitors while the *Progress* stayed in Detroit "three or four days, making additions to her equipments, after which she will proceed to Chicago."[30] Surely the decision to forbid paying visitors left Fitch Crane fit to be tied, but Captain Gifford was, ultimately, still the captain. A favorite maxim of whaling captains was that on land such men might have to listen to agents and God Almighty; but "on the other side of land, I am God Almighty."[31] Captain Gifford was not likely to listen to an erstwhile horseman so long as the *Progress* was still his vessel.

For Crane, however, there was perhaps small recompense. The arrival

Eight. Lake Ontario, 1892

in Detroit marked a slightly different but important moment in the publicity machine surrounding the *Progress*. One singular article about the *Progress* was launched from this stop and would circulate through dozens of newspapers and periodicals over the next several weeks. The reach of the text would stretch from Washington, D.C., to the deserts of Arizona and all the way to Hawaii, and would be published in periodicals as diverse as *Scientific American* and the *Chicago Journal of Commerce*. Given the size of its audience, it is worth considering the piece in-depth.[32]

"The old whaling bark 'Progress,'" the article began, "which has now reached Detroit on her way from New Bedford, Mass., to Chicago, where she and her contents will constitute for the benefit of World's Fair visitors, a complete exhibit of the whale catching industry, has a remarkable history." Despite the convoluted clauses, the article's beginning is familiar—the *Progress* would be "a complete exhibit of the whale catching industry."

The "remarkable history" mentioned in the first sentence was likewise familiar, particularly the events of 1871. However, the exaggeration of events in the article is notable. "In 1871 terrific storms scattered the fleet and all met with disaster except the 'Progress,'" suggested the text, which conveniently left out the other six whaleships that assisted with the rescue of over 1,200 people. The article also included this: "The 'Progress' which came back to New Bedford with 300 sailors, seven captains, five women and three children, the survivors of the wrecks...." This section too was littered with inaccuracies, albeit smaller ones. According to William Fish William's own account (he was, after all, one of the children rescued by the *Progress*) and later confirmed by Congressional testimony, the *Progress*'s crew and rescued souls totaled 188 people. Sometime in the intervening years, however, that number mushroomed to 226, although the specificity of the new number is odd. Was the crew being counted twice? Whatever the origin, the 226 number stuck and became standard in most accounts. So, inflating the amount yet again to 300 was really beyond the pale. Also, those saved were transported not to New Bedford but Honolulu, just as the other rescue ships had done. The fudges to the truth were not completely harmless, though. Congress just two months earlier had reopened a case dealing with the events of 1871 to address claims from the bark *Arctic*'s owners that they had been left out of the original settlement and compensation packages. The claims that the *Progress* had acted not only heroically, but also singularly, were problematic.

This article also added new details that had not been circulated before: "She has made 17 trips around Cape Horn, all of them successful. Forty times has she crossed the Arctic ocean in search of the whale and his valuable blubber." Later, the text shared with readers: "She carries six whaleboats which

107

have all seen actual service, and each one is provided with a complete equipment of paraphernalia." Both these quotes seemed to be emphasizing the authenticity of the whaleship and the whaleboats she carried. Some of this focus on authenticity might have been to counter suggestions that the whaleship didn't look like the pictures that potential customers had seen in magazines and dime novels—a charge leveled in Montreal with the quip about being short and dumpy. Similar fears might have been stirred when the *Detroit Evening News* unflatteringly suggested the ship "had the general appearance of an old-time sailing gunboat."[33]

It is also possible to probe for a deeper cultural argument behind the emphasis on authenticity when it came to *Progress*. Fitch and others were also likely sensitive to claims that attractions like replicas of Columbus' ships would eclipse a musty old whaler. The publicity from Detroit increased the focus on the *Progress* as the "real thing." As historian Miles Orvell notes, "One might imagine that the concept of authenticity begins in any society when the possibility of fraud arises."[34] Even with the Columbian Exposition's opening several months away, the promoters of the whaleship wanted to distinguish its authenticity from the many replicas, recreations, and facsimiles that would comprise so much of the Columbian Exposition, from imitation Tyrolean villages to recreations of Mount Vernon and Independence Hall, and of course, other ship-based attractions like a fake *Santa Maria*.

At the same time, it is worth mulling the question of just how much of a "real thing" the *Progress* truly was. Certainly, the ship was not a constructed replica like the *Santa Maria*, but neither was it exactly a whaleship anymore. Before she was given fresh paint and new rigging, she had been slowly decaying in harbor. The gloss of her makeover hid the realities of her age and how far she had been allowed to deteriorate. Further, the *Progress*'s lower decks had been completely converted; the decks rearranged. If told to "go awhaling" tomorrow, Captain Gifford would have been hard-pressed to do so, simply because what he captained was not a true whaleship outfitted for a multiyear voyage, no matter how much her promoters suggested otherwise. The ship was something in between, resting in a liminal space of being both a whaling vessel and an attraction for the masses.

Finally, the article is notable for what it *didn't* say. Other than the promise of a "complete exhibit of the whale catching industry," this widely-circulated piece did not detail all the harpoons, lances, and other tools used in the whaling process. Instead, the focus had shifted to the six whaleboats, which were described as "sharp at both ends and can be driven at great speed by six good oarsmen."

Why the shift in focus? Unlike the displays below deck, the whaleboats could be seen by anyone walking past the *Progress*, i.e., non-paying customers. By assuring readers that the ship and six whaleboats visible from a dock or wharf had all seen "actual service," the article was using the authenticity argument to also lure them to part with an admission fee and explore the museum within. Of course, this assumed Captain Gifford would eventually start allowing tourists back onboard.

The tow to Racine, her next stop, was one of the longest of the journey, clocking in at approximately 500 nautical miles. By now the process of preparing for a tow should have been fairly rote, as the crew had gone through several on the way to Detroit. Except this time, things did not go as planned, and the plans for the tow between Detroit and Racine deteriorated rapidly. Given the distances involved, perhaps Captain Gifford wanted to give increased attention to the particulars of this segment. Or perhaps he had grown weary of the entire process, as the Detroit *Evening News* had unsubtly suggested when it declared him "soured on strangers." And as a reporter in Buffalo had also noted, Gifford and crew had clearly developed a distaste for freshwater travel. Whatever the circumstances, what seems clear from the accounts is that Captain Gifford and crew were moving slowly, perhaps very slowly—so slowly, in fact, that a dispute erupted.

The original steamer to arrive to take the *Progress* on her journey to Racine was the *John Rugee*. According to her captain, "When I got alongside the Progress at Detroit.... I told the captain to throw me a line, but the tars, who were used to sailing on salt water where time does not count for much, were so slow about it that I came on and left them behind." The story, as reported in the *Buffalo Commercial*, would go on to describe Gifford and his crew as "old whalers" and "New Bedford relics." The captain of the *John Rugee* surmised that "they could not see that two or three hours could make any difference with me," with the article's author adding, "Time is money in lake navigation."[35]

The conflict was rife with metaphors for those that wished to see them—a tired old industry and her workers that seemed out of step and out of time with the modern, rapid pace of life and business. Of course, the entire enterprise of displaying an authentic museum dedicated to whaling was predicated upon conceiving the industry as bygone, quaint, and nostalgic. Now it was possible to see the double-edged sword of that constructed narrative—that whalers were not some tattooed trope of story books, but a caste of flesh-and-blood professionals standing in the way of modernity, efficiency, and most ironically given the name of the vessel, progress. If Fitch Crane read the article at some point, one assumes he was not pleased with how the carefully-crafted narrative of "ancient mariners" and "salty old tars" was being turned on itself. And certainly no one was pleased to see their towboat leave the *Progress* behind in a pique of frustration.

Eventually a replacement tow was found, and the journey to Racine began under the power of the steam barge *Britannic*. The captain reported that the *Progress* was "given a very cordial reception by all the boats we met on the way up from Detroit. At some points excursion boats came out to see her, and salutes from the shore and from steamer whistles were very frequent." Unfortunately, for Captain Gifford and the crew it proved to be the most eventful part of the trip for all the wrong reasons. Off Fox Island, just after rounding the tip of the Upper Peninsula, the *Britannic* and *Progress* became separated when a strong wind snapped their tow line. "We had a good deal of trouble in picking her up again," the captain reported.

Then something much worse happened. Off the shore of Milwaukee, the *Britannic* ran aground between two boulders and was stuck fast. The *Progress*, being towed behind her, still had all the natural forward momentum from the pull, and was headed straight toward the stuck steamer. Captain Gifford—a veteran of whaling since he was sixteen years old—then proved he was much more than a "New Bedford relic." He quickly disconnected his ship from the tow line so that he could turn the *Progress* towards shore, away from the stuck tug. And then, in order to prevent running aground on the beach himself, he dropped anchor to stop the *Progress*'s momentum. With the danger of serious shipwreck thwarted, he remained anchored off Milwaukee's coastline while the *Britannic* was removed from her perch by two other tugboats. The two were reunited, and the journey to Racine continued. The pair arrived into the lakeside city in the early morning hours of July 14, 1892.[36]

Although each stop since Montreal had involved some restoration of the *Progress* back to its status as a presumably complete whaling bark outfitted for an Arctic voyage *and* a museum of the whaling industry below decks, it wasn't until Racine that the crew finally had the opportunity to put it all together. In fact, as one account reported, "Scores of people hurried to the wharf and made an attempt to board and inspect the strange boat, but they were not allowed on board, the captain stating that his instructions were not to allow any person on board until the craft was fixed to receive visitors."[37] One enterprising reporter tried jumping the ship's rail in order to present his card to Captain Gifford, but got little satisfaction for the extra initiative. Although the whaling captain "greeted the reporter in a pleasant manner," the ironclad embargo on visitors remained. "The captain gave strict orders not to let any person on board," the reporter lamented, "and they were carried out to the letter."[38]

Instead of greeting tourists, the crew got to work completing the task of

Eight. Lake Ontario, 1892

creating a whaling museum. Anything packed was unpacked. Rigging that still needed to be completed was made ready. Everything was outfitted and prepared in anticipation of her big debut in Chicago on July 27. In fact, Captain Gifford told the Racine audience that they had chosen their town instead of Chicago for this phase of final instillation precisely because crowds "would have swarmed aboard in large cities."[39] In the meantime—once everyone was satisfied that "the craft was fixed to receive visitors"—the *Progress* would go on display for a week in Racine, with an additional two-day engagement in Milwaukee. An excursion boat between Milwaukee and Racine was set up for the first week. For fifty cents round trip, visitors could board the *City of Milwaukee* steamer at 2:00 p.m., visit the *Progress*, and return at 7:30 p.m.

Racine's curious got their first chance to visit the *Progress* on Thursday, July 21, although final preparations and a move to Kelley, Weeks, & Co. docks had shaved a day off the promised week. The press that day was good. The local newspaper recounted the "thousands and thousands of most beautiful and interesting articles in cases" between decks, although it erroneously gave credit to Captain Gifford for capturing and loaning the entire collection. "Lack of space alone prevents the JOURNAL recording many interesting anecdotes and extracts from the log of the ship related to its representative this morning."[40]

Those who had the opportunity to visit her in Milwaukee were likewise promised a "wonderful arctic museum" and "wonderful attraction" before she left for Chicago. In both Racine and Milwaukee there was also the added glamor of electric lighting, and for at least one evening, a small orchestra playing on deck. Captain Soren Kristiansen, a Lake Michigan schooner captain, recorded his visit in his diary on Saturday, July 23.

> She is bark rigged and they say that she is the oldest whaler afloat. Her rigging is decorated with flags of all nations and also the commercial signals, and in the night time she is lighted up by electricity. When entering her deck everything puts you in mind of that you are on board of an old timer, and coming down in her hold there is a fine and large collection of all kinds of shells, starfishes, whalebones, feathers, wings and heads of sea birds, and most everything you can think of in that line. Thousands of peoples are visiting her. Tickets 25 cents.[41]

It all seemed to be going quite well. The whaleship was within striking distance of Chicago, everything was on display, the entire attraction came across as colorful and eye-catching, and the early response was both enthusiastic and positive. But there was still one more decidedly unpleasant task to complete. With the *Progress* less than a hundred miles from her final destination, the full crew was no longer necessary. Accounts of this particular moment—and the underlying issues surrounding it—vary, but two things are quite clear: (1) the crew expected their contracts and their pay to last much longer than July 1892; and (2) they were extremely unhappy to be told that their services would no longer be required.

As the Racine newspaper told the story, the crew not only expected to be discharged in Chicago (not Racine), they also expected their contracts to last 18 months. That particular number speaks volumes about the crew's expectations. The only way it makes sense as a length of service is if the crew assumed their roles would continue through the end of the Columbian Exposition in October 1893, plus enough additional time to return the *Progress* to New Bedford when the fair was over. The *Milwaukee Sentinel* concurred, describing the scene when Captain Gifford brought up eight of the men to settle accounts at the end of their voyage. "There came near being mutiny on board the old whaleship *Progress* tonight," the newspaper reported of the night of July 21. "The men claimed they had signed articles of agreement for eighteen months before leaving New Bedford. They flatly refused to leave until the captain showed them that the articles signed were for no definite time."[42]

The *Racine Journal Times* suggested that "the sailors were wrathy [sic] to ascertain that their articles of agreement amounted to nothing and that they could be discharged at any port or place."[43] Certainly contributing to the problem was fact that English was a second language for most of the crew. Consistently described as predominantly Portuguese and Spanish, the crew now stranded on Racine's streets were likely transplants from archipelagoes like the Azores and the Cape Verde Islands. Both had been supplying the New Bedford whaling fleet with crewmen for most of the late nineteenth century. Literacy levels would have been variable, and few if any would have had a command of the language that allowed work contracts to be easily deciphered.

Even beyond questions of fairness and an understanding of their now-terminated contracts, it is hard not to sympathize with these men who were also coming to realize that their jobs were much more superficial than they originally thought. They would *not* be on the world stage of Chicago so that they could represent their friends, family, and industry back in New Bedford. Their expertise as whalers was not required or expected, a jolt to their very identities. In essence, they lost two jobs that night—the one that paid $30 per month and the one that gave them a sense of pride and self-worth. Moreover, if the *Progress*'s owners were shipping home the actual whalers of New Bedford, it begged the question: who would demonstrate the harpoons and tryworks and lances to the museum's visitors? How authentic was an "authentic whaler" without an authentic whaling crew?

As for Captain Gifford, the *Journal Times* suggested he was "satisfied" with the situation, although it should be noted this was also not uncharted

Eight. Lake Ontario, 1892

territory for a seasoned whaling captain. Whale crews frequently initiated work stoppages or made demands requiring a rereading of their contracts. "Often a reading of the ship's articles was persuasive," writes a historian of the trade. "Many men had not actually read them, or had forgotten their restrictive contents..."[44] One crew member took the news of his dismissal especially hard. "One old tar called Jack who was a little intoxicated," got particularly belligerent, eventually winding up at the police station to make his case "talking long and loud." Although there were murmurings about lawyers to end the dispute, it appears the crew members acquiesced as final preparations were made. Many remained in Racine looking for work, with the local newspaper noting that they had applied for jobs "constructing the electric street railway. They are a gentlemanly crowd and willing to do anything for an honest living."[45] The resonance of their plight echoes into today's headlines from across any number of disappearing industries.

There was one crew member, however, who did stay on, albeit in a very different role. His presence too likewise raised serious questions about the kind of "authentic" museum the *Progress* was becoming as it neared its final destination. As news accounts from the Racine and Milwaukee engagements detailed the visitor experience, a brand-new reason to visit the *Progress* began to receive occasional mention: Jimmy Kanaka [Kanaca], "an ex-king of the Fee Jee Islands." Later when the *Progress* was in Chicago his presence would be described as: "A real Fiji Island King is on board, and is the only one who has visited Chicago."[46] The sudden appearance of royalty aboard the *Progress* suspiciously coincided with the departure of her multiethnic crew.

"James Kanaka" had appeared on the crew list from the beginning, his name serving as a calling card to his ethnic heritage. "Kanaka" was a catch-all term for South Islanders, Fiji included. Writes whaling historian Everett Allen, "If a shipmaster could not spell a native's last name, he often recorded him in the ship's articles as John, Sam, Joe, or Henry Kanaka..."[47] In fact, the name "Jimmy Kanaka" would later be used by playwright Eugene O'Neill (who favored maritime themes in his work) for one of his stock, non-white villains, a harpooner "who makes his entrance dressed like Tarzan, wearing only a loincloth and sheath-knife."[48] And the *Progress*'s Jimmy Kanaka had another trait that made him perfect for a little playacting: "the South Sea Islander was conceived as a great asset, since he was tattooed almost from his head to his feet."[49] Whether Jimmy himself developed the scheme or it was suggested by others, he used his physical appearance to reinvent himself into this rather dubious character of "a real Fiji Island King" and "the only one who has visited Chicago." Other crew members were not so lucky. Family back in New Bedford had to enlist the help of the Racine Police Department to aid in tracking down crewmember W.J. Carter. He was found destitute at a Racine boarding house and had been digging sewers to earn enough money to survive.[50]

But beneath all the activity, all the visits, and all the excitement of the *Progress* finally so close to arriving in Chicago, there were seeds of doubt. It had been easy to spot them along the way, but now that the *Progress* had arrived at Chicago's doorstep in Racine, the press had the opportunity both to reflect and to sharpen their knives. "While the Progress is a whaler she is not the great curiosity that the owners would make her," wrote the Racine reporter who jumped the ship's rail that first day of her arrival. Perhaps his jab was simply sour grapes at being kicked off the ship, but his was not the only such comment.[51] "It is bad," wrote Chicago's own *Daily Inter Ocean*, "for the syndicate which brought the old boat up from salt water that she can not be put in a building or under a tent. There is far more curiosity to see a real old-fashioned whaling ship itself than the museum of whaling appliances she may have on board." In two sentences, and with Cassandra-like accuracy, the author cut to the heart of the matter. But it would take the next year and a half in Chicago to fully answer the question: Did people really want to experience a museum dedicated to the dying industry of whaling?

Nine

Chicago, 1892

Steele MacKaye had better things to do with his time. His envisioned 10,000-seat theater—the Spectatorium—was reaching a critical juncture, with the sale of stock in the Columbian Celebration Company about to begin. MacKaye had originally hoped to be given a piece of land within the fair's boundaries to build his massive theater and indoor reservoir for Columbus' musically-interpreted crossing. That request had been turned down. It meant the penniless MacKaye now had the unexpected price tag of Chicago real estate—adjacent to the fairgrounds no less—to add to the list of expenses that seemed to be exploding before his eyes, quickly approaching a million dollars.[1] He desperately needed investors. MacKaye had thousands of details to attend to; instead, here he was on a steamer in Lake Michigan watching wagonloads of food and drink from Kinsley's—"The Delmonico's of Chicago"—carried aboard by a seemingly endless stream of African American waiters.[2]

When the *Adrienne* got back underway the lake breeze began to cut the blistering temperatures from the heat wave that had gripped the city. Music by a mandolin orchestra was wafting from the bow. MacKaye looked around and reminded himself why he was here. The entire day had been arranged by millionaire coal baron Henry Weaver—a man that MacKaye particularly wanted to keep happy and close. And Weaver had also managed to secure the attendance of MacKaye's most important advocate and potential business partner, Benjamin Butterworth, Secretary of the Fair. Butterworth was a powerful man, widely recognized as one of the driving forces behind the Columbian Exposition. He was also someone MacKaye wanted to keep happy and close. So, Steele MacKaye got work mingling, talking, and shaking hands. Besides, it would make a good story in the end … how many people could say they had ridden a whaleship to Chicago?

July 27, 1892, was Henry Weaver's big day. Not only was the *Progress* arriving in Chicago, it was doing so as a major event. Multiple newspapers had reporters on board, and the guest list was sufficiently impressive, with Benjamin Butterworth as the most notable and prominent attendee. Weaver had almost gotten Mayor Washburne to join them, but the city's leader was

The Last Voyage of the Whaling Bark *Progress*

only able to see the party off from their Chicago pier. Still, Cook County Sheriff James Gilbert was aboard, as was the city comptroller Horatio May. Egbert Gillett, the Magic Yeast mogul, joined the party. Real estate developer George Stodder, who had recently opened the Stodder Building near the fairgrounds, was there, as was Anson Gorton, an alderman and the manager/general agent for Wells Fargo in Chicago. Commissioner of Public Works J. Frank Aldrich was onboard, likely working the crowd ahead of his run for the U.S. House of Representatives. Aldrich had also signed on as a stockholder in Weaver's Arctic Whaling Exhibit Company. Relatives likewise joined Weaver for his big day, such as members of his wife's Guthrie family including Ossian Guthrie, who had been part of the original delegation to New Bedford.[3]

As the steamer headed north, the *Progress* was being towed south, "with all her sails set and her flags and streamers flying gayly in the breeze." The two met off the coast of Fort Sheridan, with the *Herald* reporting that upon seeing the fully rigged bark "every landlubber gave his trousers a hitch and assumed an air intended to be old salt like." Once the two vessels were aligned, the transfer of the several dozen prominent Chicago businessmen and city leaders began. The whole spectacle of bringing the men aboard was vaguely reminiscent of the transfer of whalers onto the very same whaleship in the Arctic over two decades earlier.[4] The parallel was reinforced when the delegation was fully assembled on the decks. There the men were presented to none other than Captain James Dowden, former captain of the *Progress* and famous hero of the 1871 disaster. Henry Weaver had managed to secure a star for his floating museum, who stood alongside the much less well-known Captain Gifford as the voyage to Chicago got back underway.

The reveal of Captain Dowden was not entirely a surprise to those that had been paying attention. Dowden had resigned from the New Bedford police force where he had been serving for the last few years. Weaver had brought him to Chicago several days earlier and he had inspected the ship in Racine before this big debut.[5] He had been staying in Chicago's upscale Tremont House awaiting his formal reunion with the *Progress* and had given at least one interview. The *Daily Inter Ocean* described him in the opening paragraph: "He's an old man now, but though too wide of girth to spin up a spar or dangle on a rope, he is sound of limb and gruff of voice and jolly as the typical tar that he is." The addition of Dowden to the attraction placed further emphasis on the dramatic events of 1871. Although the rescue had consistently been mentioned during the previous stops leading to Chicago, Dowden was now available to elevate the narrative and could provide firsthand accounts of the events. Lest visitors forget or miss the connection, Dowden wore "an

Nine. Chicago, 1892

THE ARCTIC WHALER PROGRESS.

"The Arctic Whaler Progress." G.A. Coffin. Here the whaleship is receiving guests on Lake Michigan. The size of the *Progress* can be appreciated as the shuttle carrying Henry Weaver and his guests sits dwarfed alongside the whaler. The Chicagoans must have recognized the uniqueness of this private invitation to ride the *Progress* into the Windy City. From *"There She Blows"* (Chicago: Arctic Whaling Exhibit Co., 1893).

immense gold chain from which is suspended a large gold medal given by the 226 whom he rescued in the North Seas."[6]

In addition to being a flesh-and-blood connection to one of whaling's most famous incidents, Dowden represented something else—an archetype. Piece after piece emphasized Dowden as fitting the role that was expected of him as a spinner of yarns and stories, because in the words of one article "that's the kind of man Capt. Dowden is." The accounts were verbose and exuberant on this point. In the words of one: "he will stand on her deck during the World's Fair and tell in his interesting, blunt, sea-dog fashion the history of the wonderful craft." Another: "the party surrounded Capt. Dowden and listened to the old veteran while he spun yarns about whales and harpoons and kyacks [sic] to their hearts' content." And yet another: "This is something new for the Captain and he has not attained the fluency that experience will give him, but it is all the better for its bluntness and lack of professionalism."[7]

However, just as had happened with Captain Gifford on the way to Chicago, these accounts represented the double-edged sword of presenting the men of whaling as all "old tars" and "sea dogs"—they remained relegated to a status that denied them their professionalism and place in the modern world. This was happening in spite of the fact that whaling continued throughout 1892, with over 50 voyages launched from ports all around the United States, New Bedford and San Francisco especially. In fact, 1892 went down as an excellent year for the whaling industry. The *Boston Daily Globe* reported "Although the whale fishery of this country is not what it once was, it still assumes quite large proportions, gives employment to several hundred seaman, and is still credited with fairly large number of vessels. As a financial venture it has been more successful the past year than for several years."[8]

This was one of the central conundrums of trying to memorialize an industry that technically was still active and alive. The Massachusetts Board of Managers for the Columbian Exposition argued just such a point in their commemorative *Massachusetts of To-Day* booklet for the fair: "The whaling industry, in which the city's wealth was earned, while sadly reduced, is by no means extinct; and New Bedford now as in the olden days, does the largest whaling business of any place in the world."[9] When the New Bedford Board of Trade, the same entity which sold the *Progress* in the first place, published a town history a few years earlier in 1889, they followed a similar tack, in which whaling was carefully couched in language of both history and optimism: "Despite its decline, it is an undisputed fact that the whale-fishery of our city, as now carried on, has a larger interest than the whaling of the whole outside world together."[10]

This perhaps explains why Captain Gifford was not staying with the *Progress* and Captain James Dowden was. In fact, Captain Gifford had been formally discharged at the same time as the other crew, although he had remained aboard for the pomp and circumstance of this arrival day. Now the ship was Dowden's, and not only did Dowden have the personal connection to the *Progress*'s place in the dramatic stories from 1871, he also was no longer

Nine. Chicago, 1892

an active whaler. As if preserved in amber, Dowden stood outside of the present day, able to continue a mythology of whaling as an ancient art, relegated to the distant past. This, of course, was not true, but it needed to *seem* true for the museum venture to work. When the party of Chicagoans stepped onto the *Progress* and met Captain Dowden, they needed to be transported by the stagecraft of the whaleship and the performance of her lead actor. And by all accounts they were—a fact that surely was not lost on America's most famous theater producer who happened to be in the audience, Steele MacKaye.

The crowd of men began their tour of the ship, particularly the museum between decks. Again, just as Captain Dowden's appearance should not have been a surprise to an astute reader of the daily newspapers, the structure and content of the *Progress*'s displays had also been previously revealed for anyone keeping abreast of the whaleship's press coverage. Her promoters had secured a lengthy piece in the Sunday edition *Chicago Tribune* prior to her arrival in Chicago, complete with detailed illustrations and verbose descriptions spanning two and half columns. It is a fascinating read because it serves as one of the most complete descriptions of the exhibit as it was conceived heading to the Windy City.

Much of the article covered familiar territory—the origins of the whaleship, her journey to Chicago, statistics about her productivity as a whaler, and the events of 1871. The article also reasserted the dominance of Henry Weaver in the narrative: "The ship was purchased by Henry E. Weaver of Weaver & Getz, and by him sold to a stock company of which he is the chief member." There is no mention of Fitch Crane or the suggestion that Crane originated the idea of purchasing a whaler for display at the Columbian Exposition. The article also took back one other claim of Crane's. According to the piece, it was Captain Dowden who scoured the New Bedford area for artifacts to display, not the young Chicagoan.[11]

The piece then began the reader's tour of the vessel. The tryworks were introduced and explained, their on-deck housing structure described as reminiscent of the "boiling-house found in maple sugar groves." Given the notoriety of tryworks' smoke and smell, it was a generous comparison. But what followed a visit to the tryworks is illuminating—an extensive description of the ship itself. The prose is comprehensive, bordering on exhaustive: the anchoring system, the masts and rigging, the deck-house, the captain's quarters, the skylight, even the sofas and dining table. Although perhaps this was to be expected for a feature article about a ship, looking closer one realizes what is missing. For nearly 300 words and multiple paragraphs, the author(s) studiously avoided any mention of whaling or the purpose of the vessel they were so thoroughly describing. Yes, the ship was being recreated through words, but in largely generic terms of sailors, cooks, and captains that could apply to any number of maritime vessels.

Only in the third and final column of the article do readers break away from more generic nautical descriptions and return to what was unique about the whaler. Except the narrative did not return to whaling implements, whaleboats, or the blubber room. Instead the article led vicarious visitors to "the space between decks, including the cabin and space used for cargo." It was this space that was being "turned into a museum for treasures of the deep. Immense swordfish blades, jaws of whales, whales' teeth, turtles seven feet long, all manner of marine curiosities, sponges, shells, corals, seaweed, and curious sea growths crowd the showcases which will occupy this space."

This too is revealing because it has so little to do with the actual practice of whaling. Outside of jawbones and teeth, whales are not even mentioned. Instead, the reader is left to imagine "marine curiosities" including sponges, shells, coral and seaweed—items that have no relevance to the whaling trade. Lest one assume the writer simply had a misplaced focus, it is worth comparing this description to what Captain Soren Kristiansen wrote in his diary after his visit to the *Progress* in Milwaukee just a few days earlier: "a fine and large collection of all kinds of shells, starfishes, whalebones, feathers, wings and heads of sea birds, and most everything you can think of in that line."[12] Both the article for widespread readership and the private thoughts of a nautical diarist turn out to be quite similar.

The turn of phrase "marine curiosities" would prove vitally important and prescient. All throughout the first half of 1892, the *Progress* had been conceived as a testament to the industry of whaling. Press and publicity emphasized her "every variety of implement used in the chasing of sea monsters"; "all the inventory of a whaling voyage"; and "quite a museum of articles used and obtained by whalemen," among many other similar sentiments.[13] There had been stray mentions of polar bear skins or similar deviations from the business of whaling, but by and large, the objects mentioned throughout her pre–Chicago publicity had been by and from whalers—harpoons, lances, logbooks, photographs, etc. Now as the *Progress* edged closer to her Chicago debut, the narrative was changing for both the syndicate and the viewer. Shells, feathers, corals and seaweed were rising to the forefront.

A comparison of the illustrations between the *Montreal Daily Witness* feature at the beginning of the *Progress*'s voyage and the *Chicago Tribune* feature at the end is illuminative in exploring this change of narrative. Both provided drawings of the *Progress* at dock. However, the Montreal paper illustrated its article with renderings of four whaling instruments, each of which was carefully numbered and then fully described in detail within the text.

Nine. Chicago, 1892

The Chicago paper, by contrast, showed harpoons, guns, and lances in an almost bouquet-like arrangement, with the accompanying title "some curios." Another Chicago illustration gives prominence to the giant sea turtle. All of this would seem to suggest that the earlier Montreal feature worked under the assumption that the exhibit's value was that the intricacy of whaling would be explained using authentic artifacts. This is exactly what the originators in New Bedford had intended. The latter feature began to reveal the cracks in that veneer—that perhaps people would be more interested in "marine curiosities" than the intricacies of whaling.

Only at the very end of the *Chicago Tribune* article does the whaling-centric narrative begin to reassert itself. There was the promise "to have published a short description of the whaling business, which will be furnished to the ship's visitors." Once on display, the crew would also mount chases of imaginary whales to illustrate the practice of whaling and the tools used. "It is proposed to have practical whalers at hand to show visitors every appliance connected with the business and explain its use," the article stated in its last 100 words of the piece. Still, we know now the dubiousness of this claim given that the "practical whalers" had already been discharged and were looking for work back in Racine so they could afford to get home to New Bedford.

The "Whales at the Fair" feature came out on Sunday, three days before Henry Weaver's carefully orchestrated occasion. Yet accounts of what Steele MacKaye and the other handpicked visitors on July 27 experienced on their tour of the ship were equally revealing in this tonal shift. From the *Herald*: "They saw an alligator turtle seven feet long; the flipper of a porpoise looking like a petrified human hand; whales' teeth of a thousand shapes and sizes; sea horses; costumes worn by members of the Greely relief expedition in the frozen seas," and so on through a litany of non-whaling items ranging from a quadrant purportedly from the *Mayflower* to native dresses from the South Seas to a mummified boy from Australia. Only after this exhaustive list was there mention of "a mighty array of harpoons and whale-killing weapons." From the *Inter Ocean* there is a similar pattern: "Trophies are everywhere, on the deck, in the hold and in the cabin. Cases of bones, mammoth turtles, sharks' jaws, skeletons of whales and other interesting parts, swordfish and suggestively the celling is studded with stuffed star fish." Only after this list is there mention of harpoons and "bum" guns for the actual work of whaling.[14] The *Times* offered a rare exception, describing the "harpoons in readiness, the whale boats in position to be launched ... the boilers to try out the blubber." However, this may have been because the reporter was enamored with the *Progress*'s Arctic connections in the midst of the city's scorching heat wave. Under the headline "Fanned by Arctic Breezes," the reporter's willingness to talk whaling likely stemmed from his desire to "think hard that your weather is very cold. Then look around you and you can easily imagine yourself on a whaling expedition to the Arctic regions."[15]

Arctic-bound flights of fancy notwithstanding, why were accounts of the *Progress* suddenly so different, compared to earlier descriptions and explanations? And perhaps more importantly, was this a conscious and deliberate attempt to change the narrative about the *Progress* by Weaver and his associates? One explanation is that all the starfish and shells and mummies and grass skirts had simply been in storage until the *Progress* reached Racine; therefore, it took until mid–July before people got a real look at the complete exhibition. Prior to then the main (or only) things to look at actually were the instruments of the whaling industry. After a complete outfitting at Racine, people saw the complete assemblage of "marine curiosities" and therefore the descriptions naturally shifted.

However, it is impossible to ignore the fact that Henry Weaver had nearly two months to think about his purchase as the *Progress* made her way toward Chicago, including reports back from Fitch Crane about how people (and the press) were reacting to what they saw. And in that time, he had clearly been talking more and more with Steele MacKaye, so much so that he invited MacKaye to join the whaleship's arrival party. How much MacKaye actually cared about whaling is an open question. He had been to New Bedford at least once on his 1875 Lyceum lecture tour for "The Mystery of Emotion," which promoted MacKaye's scientific approach to dramatic expression on the stage.[16]

What MacKaye *did* care about was stagecraft and the mechanics of a good show. He now saw his Spectatorium in Chicago as the ultimate triumph of art and aesthetics meeting science and invention. "The aim of this co-operation is to develop a form of entertainment which shall be the most fascinating and impressive while at the same time the most noble and beneficent of any that has yet moved the minds of mankind."[17] Weaver must have realized there was only so much he could do with the *Progress* to achieve the higher level of drama and spectacle that MacKaye so easily pitched. Still, as he and MacKaye compared notes about their respective enterprises, Weaver must have also realized that he was under no obligation to stick to New Bedford's original vision for an authentic and comprehensive exhibit of the whaling industry. In words that MacKaye himself would completely understand, Weaver was learning that sometimes you need to change the script.

This evolution of thought can be seen in two advertisements for the *Progress* in the weeks after her Chicago entrance. The first, placed immediately after she arrived and went on display, encouraged readers to visit "The

Nine. Chicago, 1892

Old Historical Whaling Bark Progress and Museum." The second, barely two weeks later, urged instead: "Take an Hour and Visit the Great Museum of Marine Curios." The first ad mentioned that the curios were "brought home by old-time whalers"; the second excises any mention of "old-time whalers" or their connection to the objects on display. Where the first ad places "and Museum" under the words "Whaling Bark Progress," the second ad places the "Great Museum of Marine Curios" first.[18]

Back at the shipboard party on July 27, things were winding down as the *Progress* anchored at the Government Pier at the end of Van Buren Street. Here Weaver seemed to split the difference between these two notions of what kind of attraction the *Progress* was. His new arrival to Chicago slid beneath a banner that read: "Arctic Whaling Museum: 10,000 Marine Curiosities Between Decks.," effectively ticking both boxes of whaling and curios. Still, whatever ties the *Progress* had to New Bedford's Board of Trade and their vision of an authentic representation of whaling were now dissolving as the ship settled under that "Marine Curiosities" banner. It was a point emphasized by Henry Weaver's cousin who wrote a letter back home to his local newspaper describing his relative's handling of the museum idea: "What New Bedford people failed to do has been accomplished by an enterprising gentleman from the west—that of fitting a whale ship for exhibition purposes at the World's fair in Chicago."[19]

Before Steele MacKaye and the several dozen other Chicagoans departed from the whaler, Benjamin Butterworth—the man many credited with making the Columbian Exposition in Chicago possible—made a few concluding remarks that reaffirmed the *Progress* now belonged here on Lake Michigan, not the wharves of New Bedford. "It is to just such public-spirited citizens as Henry Weaver that Chicago owes its unrivaled prosperity. We see here today a reminder of the commerce of a past age, and through the pluck of those business men of Chicago are we permitted to see one of the ships with which this industry was carried on."[20]

The other concluding words came from the north the next evening in the *Milwaukee Herald*: "Judging from the morning's papers, Chicago professes to be amazed and shocked by the imaginative magnitude of the whaling stories told by good Capt. Dowden of the whaler Progress. It is nothing more nor less than a monstrous piece of cheek for Chicago to be startled at the spinning of any yarn—no matter how windy."[21]

With the *Progress* safely in Chicago, Weaver's first order of business was finding a better location. The Government Pier was lakeside, which attracted those seeking relief from the brutal city heat, but otherwise, the venue was less than ideal. It was out of the way from the city's center with the route bisected by heavy train traffic. It was a destination known more for fishermen than sightseers. Weaver began work getting a better berth right in the heart of Chicago. By early August he had negotiated a spot from the Grand and Norton Line, the same transportation company that had provided excursions between Milwaukee and Racine just a few weeks earlier. The prime spot was located on the Chicago River at the foot of the State Street Bridge, and the whaleship was in its new home by the first week of August.[22]

The move was enough to generate another small round of press coverage. An article in the *Chicago Mail*, while flattering, barely even acknowledged the *Progress* as a whaling ship. Instead, it suggested that a square-rigged vessel was "something seldom seen in this harbor," and that the exhibition of "marine and island curios … the handiwork of the Esquimaux of the arctics and the cannibals of the tropics" was "one of the most attractive museums that has ever been brought to the United States."[23] The *Tribune* also ran another short piece on the attraction, starting with the subhead: "Interesting Exhibition of Marine Curiosities and a Tar of Ancient Days." The ship was described as an "old whaling vessel … which sailed the north and south seas back in the '40s." It went on to applaud the enterprising spirit of those who had brought her to Chicago. "Yesterday it was thrown open to the public," the article continued "and the dear people allowed to inspect its 'museum of marine curiosities' at a fixed sum a head." As before, the objects described focused far more on the "things mariners used to pick up," such as shells, idols, weapons, swordfish swords. "Apparatus for killing the huge mammals" is the only material reference to whaling.

The article also spent significant ink describing "an old tar" who sat on the forecastle deck. Although not mentioned by name, it seems certain that this was Captain Dowden, given this was the first few days in the new State Street location and the most likely time for Henry Weaver to capitalize on the crowds and attention with his hired, authentic whaling captain. "He was a trifle better dressed than the old tar of history, but he chewed tobacco just as vigorously and told the crowd around him many a tale…. A reporter for THE TRIBUNE was pained to notice that he did not say 'shiver my timbers' once, even in the most exciting periods of his stories, but otherwise he was all the most critical of dime-novel readers could desire."[24]

By now both halves of this article were familiar. In the first half one sees the description of a "marine curiosities" museum with nominal connections to the practice of whaling. Further, this article continues to advance the idea that these objects were collected by whalers and brought home by whalers from distant shores. This was also woven into the advertising from the *Progress*'s first few days on display. It is worth considering how this idea of mari-

Nine. Chicago, 1892

ners picking things up and collecting them might have been understood and internalized by those visiting the ship, because it ultimately worked to obviate the realities about whaling and the kinds of men who made up most of the industry's ranks.

The truth was that whalers spent very little time in their three- to five-year voyages on shore. And when they did, few rank and file crewmen took their precious hours of shore leave to collect cultural mementos for domestic display back home. In the words of one historian, "Whaling seamen on shore leave were a boisterous, pleasure-seeking rabble, whose ideas of pleasure were not over-refined, too frequently their tastes rose no higher than the grogshop and the brothel...."[25] Of course, this did not apply to the entire crew, and officers especially could have fit the profile of these artifact-collecting denizens of America's whaling fleet.

We know from various earlier press accounts that the objects described were indeed gathered from New Bedford families that had brought items from around the world back into their parlors and offices. But such collectors were predominately the captains, officers, and masters of whaling voyages who only made up a tiny fraction of the fraternity of whalers. In fact, when New Bedford first announced the *Progress*'s sale it was assumed the objects displayed would largely come from "merchants in this city" as well as the *Progress*'s owners.[26]

The difference between whaling elites and the hardscrabble hands who made up most crews was lost in the claim that the objects on display were an accurate depiction of whaling and whalers. It effectively took whalers—who lived in a dominantly male and often violent world—and turned them into servants of domesticity and home décor. Further, these totems of "shells beautiful and rare" conveniently replaced for museum visitors the actual goal of whalers, which was bloody, raw strips of flesh ripped from a recently-living creature. The idea of whalers (and actually, not even "whalers" in this most recent article, but the much more bland moniker of "mariners") picking up objects from around the world took whaling and made it into something that was easily and uncomplicatedly consumable by viewing audiences.

In many ways, the *Progress*'s connection to the story of rescue in the Arctic in 1871 served much the same function. This had been one piece of coverage that was consistent throughout the *Progress*'s tenure thus far—there was almost always some sort of reminder or retelling of Dowden's contribution to the rescue. Yet when we stop and think about what else this story was doing, besides spinning an entertaining yarn, it becomes apparent that this was another way of discussing whaling without delving into its gruesome and violent realities. The accounts often mentioned the women and children who found refuge on the *Progress*, softening the androcentric world of the whaling trade. The story also served as a good morality tale, emphasizing sacrifice over profit and selflessness over personal gain. Dowden's gold medal and chain were often mentioned, a visual, material reward for his virtue. And while all of this was certainly true, it was hardly an authentic representation of how the whaling industry worked and function year after year, decade after decade. In fact, the story was a fantastic way to *not* talk about whaling, since one of the central themes of the narrative was that whaling was abandoned in order to save the 1,200 souls. Visitors could walk away from this portion of their tour aboard the *Progress* convinced they had engaged in an authentic whaling experience in which no whales were actually killed!

The focus on the 1871 story also leads naturally into the second half of the article, which was also now familiar for its treatment of the *Progress*'s human actors and interpreters—Captain Dowden and the crew. Even more than the objects, the people associated with the *Progress* were seen through a lens of story books, whaling tales, and dime novels, the latter explicitly cited by the *Tribune* which suggested that even the most critical dime novel readers would come away satisfied by the men they met. Again, the description places Captain Dowden in a presumably bygone time, separating him from 1892 Chicago and America so that visitors can effectively be transported. Like so many other elements of the visitor experience, Dowden himself has been cleaned up and literally sanitized. His clothes are now described as better "than the old tar of history," assuring readers that no hint of blood or gore will be spotted and offend. Instead, like a good costume, Dowden's clothes help propel the story and the visitor out of their modern lives and into the pages of a rousing maritime adventure.

Thus, Dowden's purpose was clearly seen in the article as entertainment over education. The fact that so many articles reveled in his telling of tales, yarns, and stories is an indication of this priority. The *Herald*'s feature on the *Progress*'s July 27 arrival was particularly prone to this, including an illustration labeled "Steele MacKaye swallows a whale story." The *Tribune* piece several days later, after the move to State Street, continues in this vein, highlighting that Dowden "told the crowd around him many a tale of 'when we were in the South Seas in '52' and 'the time we went 'round the Horn in '58.'"[27]

One additional implication in labeling these narratives as "whale stories" and "tales" was that they were assumed to be part truth and part fiction. It was left to the audience to determine in what quantities truth and fiction made up each story. Just as P.T. Barnum trafficked in displays of both authentic attractions and "humbugs" a generation earlier, so too was Dowden—presumably at Henry Weaver's instructions—crafting an attraction based on certain "arts of deception," with visitors listening to an old whaler spin tall tales and whale stories, and leaving it to them to puzzle out what was real and what was fiction.[28]

All of this was a far cry from a pedagogical and authentic representation of the practice of whaling that had been articulated in New Bedford just a few months earlier. But was it more in keeping with what actual, paying customers wanted? As this visitor experience comes into focus it seems to suggest what audiences demanded (or at least subconsciously desired), which was for whaling and the *Progress* to fit into a vague, fuzzy understanding built up from reading adventure books, dime novels and serialized stories. The fact that the author inserted the phrase "shiver my timber"—derived from pirate stories and lore, including the recently published *Treasure Island*—into an article about whaling shows just how nebulous this understanding was. Just as whalemen had been converted to the vaguer "mariners" earlier in the piece, now the *Progress*'s authentic whaling captain was collapsed into a mélange of maritime tropes covering whalers, pirates, polar explorers, and any other "salty tar" the visitor might have encountered in print or popular culture.

It is also important to consider how performative Dowden's task truly was. He stood or sat on the forecastle deck, a bit of stagecraft Steele MacKaye surely would have recognized because the Spectatorium's Christopher Columbus was scripted to stand in *exactly the same spot* of his "Santa Maria" for the climax of Act Two.[29] As already noted, Dowden's cleaned-up clothes were seen as a costume, only a "trifle" removed from the "old tars of history," and his chewing tobacco habit became an appropriate character detail. In this way, the *Progress*'s authenticity for presenting the whaling profession gives way to something else—authentic props for Captain Dowden himself. All the early ideas of lowering whaleboats, chasing imaginary whales, or executing authentic whaling maneuvers are nowhere to be found, in part because the crowded waterways around the State Street Bridge would not allow such demonstrations. Therefore, in the absence of any other narratives, the ropes, the spars, the whaleboats, and so on simply become idle props designed to reinforce Dowden's own legitimacy as a "real" sea dog.

Finally, just as all these narratives worked to soften and obviate the violence and filth of whaling, so too did they shield visitors from whaling's exploitative practices. By the 1890s the lowest members of the crew were almost all men of color, while officers were still most frequently white. In this way, whaling hewed suspiciously close to slavery. These crewmen would often find themselves with little money or no money to show for their years of service, while others actually discovered themselves in debt by voyage's end. The complicated system of "lays," or pay at the end of voyage based on a percentage of profits, was stacked to favor owners and officers. Working and living conditions were awful, leading to countless desertions, work stoppages, and more than a few outright mutinies. One need only look at the disposability of the crew in Racine is to be reminded of these harsh realities.

Of course, none of this would be acknowledged aboard the *Progress*. The filthy forecastle where crew slept amid roaches and rats had been made appropriate for middle class tourists to wander past cases of marine curiosities. The strict hierarchy and discipline of an authoritarian regime was subsumed into a pastiche of yarn-spinning, kindly old tars. The fable of jaunty whalemen who just wanted to tell tales to visitors closely resembled the similarly-spun narratives from the South of "happy negroes" content to work on paternalistic plantations. All the sharp, harsh, bony edges of whaling's true form had been smoothed over into a comfortable and sellable version of reality. This included the fact that the original crew—men of color from foreign island communities—had been discharged in Racine and replaced with local, freshwater deckhands. The record is mute as to whether these new actors were black or white, but given all the other attention to a "clean" and "safe" version of whaling, it is not difficult to make an educated guess for the latter.

All of this was thoroughly reinforced by another publication: the four-page souvenir program developed for the exhibit. The impressive cover featured the *Progress* with its full complement of international flags and banners, sitting before a vague landscape of mountains or icebergs set in a circle. The text announced "The Arctic Whaler Progress A Complete Marine and Whaling Museum. On Exhibition at State Street Bridge from 9 A.M. to 10 P.M. Admission 25 Cents." This document is what the *Tribune* had referenced back in July when it promised that there would be a brochure "to have published a short description of the whaling business, which will be furnished to the ship's visitors." However, if even a short description of the whaling business is what readers were expecting, they would have been sorely disappointed.

"Every one has read of whalers and their wonderful adventures in quest of those monsters of the deep, but how many have seen a ship and crew who have braved the perils of the deep that ladies might have the indispensable whalebones for their dresses and corsets." The opening paragraph of the piece once again establishes the pattern of referencing tales and previously-read

Opposite: **Cover, souvenir brochure, State Street Bridge, Chicago. 1892. Designed for the *Progress*'s State Street engagement before she was moved to the fairgrounds, this souvenir brochure helped shift the focus of the museum away from authentic whaling and towards topics such as arctic adventures and assorted, exotic curios gathered from around the world (courtesy New Bedford Whaling Museum).**

Nine. Chicago, 1892

adventures as an assumed baseline for understanding and appreciating whaling. It then turns its attention to objects of whaling. "Amidship, on the main deck, are the try work, huge furnaces for melting the blubber from the whale and making it into oil. On the sides of the ship are the whale boats used in killing these huge creatures. Scattered around are the lances, harpoons, bomb-lances, guns, etc. used in prosecution of their calling." These 50-some words are the extent of the details. That is all there is that one can point to as the complete description of the business of whaling. And even these paltry sentences are somewhat dismissive, such as the suggestion that the authentic implements of whaling are simply "scattered around," something that never would happen on an actual whaler. Then the very next sentence is set in italics to show its importance and informs readers that below the deck *"is a museum, the like of which has never been seen in Chicago, and objects here shown run into the thousands."*

The message is clear. Having dispensed with the business of whaling as quickly as possible, readers' attention is turned to the museum of marine curiosities. Some of these objects had whaling connections, such as the baleen ("whiskers" according to the brochure) and whale jaws. But the vast majority did not. Spanning ethnography, botany, conchology, and more, the list was becoming increasingly familiar—the mummified Australian boy, the *Mayflower* quadrant, the costumes from both the Arctic realms and the South Seas. It is also here that Jimmy Kanaka received his nod as "a real Fiji Island King is on board and is the only one who has yet visited Chicago." The text then shifted to the requisite retelling of the Arctic disaster, with a particular focus on Dowden, who "gave up his state-room to the ladies, and did all in his power to render their voyage comfortable." Mentions of his gold medal and gold watch of appreciation were also included. It concluded by telling readers of the *Progress*'s trip to Chicago and how she was now in the command of Captain Dowden, and manned with "a crew of sturdy blue-jackets, old sailors, who have cruised in all parts of the world." In the final two sentences readers are assured they have made the right choice in visiting: "It was the intention of the owners to bring the ship here for exhibition at the World's Fair, but realizing the crowds that would visit her at that time, they wisely decided to give the residents of Chicago the first chance to see this venerable relic. That they have made no mistake is shown by the large number of daily visitors."[30]

By now the dichotomy between what had been originally conceived and what the *Progress* had become was crystal clear. But it is worth pausing here to consider alternative narratives. It is easy enough to criticize the rush past the business of whaling to focus on the sensational and eclectic nature of the marine curios; however, was it possible to expect more from the museum?

Nine. Chicago, 1892

Were there examples that could have guided Weaver, Crane, and others in a different direction? In fact, there were. The National Museum's (aka Smithsonian Institution) exhibit of whaling at the 1883 Great International Fisheries Exhibition in London had been staged only a few years earlier and then was largely repeated the next year in two American venues—Louisville and New Orleans.

That display had emphasized an encyclopedic fidelity to whaling. The harpoon section, for example, was designed as a four-part "exhaustive treatise on the past and present methods of the capture of the whale adopted by all nations that have participated in this fishery." A section entitled "Implements Used Exclusively on the Vessel" was described as an immense pyramid that not only accurately displayed spades, knives, gaffs, pikes, forks, and bailers but also triggered an olfactory response: "About the top of this immense structure of whaling apparatus, which is strongly suggestive of the odor peculiar to a whaling vessel, the boat-waifs for locating dead whales are placed in prominent positions."[31]

We might wish to forgive the Chicago syndicate for doing as well as they could and ultimately just giving audiences what they wanted. Perhaps it is tempting to assume that they could not do any better than having some harpoons and lances "scattered about." But this ignores the history of whaling displays prior to the *Progress*—displays that *did* emphasize completeness and accuracy, right down to embracing a whaleship's peculiar odor. Deviating from this model, and the possibility of doing more than briskly pointing out a few whaling tools before ushering visitors down into the "main" feature of curiosities, was both a conscious and deliberate choice on the part of the exhibitors.

Weaver's choices, when more deeply considered, were also surprisingly anachronistic. As the Smithsonian exhibit revealed, the modern trends of the era emphasized classification and organization. The Gilded Age was awash in attempts to arrange the world into discrete and exhaustive departments and categories. It was a cultural "search for order" that underpinned everything from department stores to museums to universities. In fact, the world's fair itself (the *Progress*'s ultimate destination) was an exercise in organization writ large—an attempt to catalog the world through an exhaustive schema. The Smithsonian's comprehensive and thorough exhibit was very modern in this sense, with all the harpoons together, then all the lances, then all the knives, etc.

Weaver's "museum of marine curiosities," by contrast, was much more akin to the museums of a bygone era. The higgledy-piggledy mixing of whale bones and shells with mummies and grass skirts harkened back to Ameri-

The Last Voyage of the Whaling Bark *Progress*

ca's earliest museums, such as Scudder's American Museum in early nineteenth-century New York. Scudder too displayed a whale jaw, mixed in with an "Indian Mummy, dug out of a saltpetre cave in Kentucky," some mammoth teeth, and a "group of large hairballs."[32] P.T. Barnum was perhaps the most famous purveyor of this sort of hodgepodge, freely mixing the realms of natural history with "living statuary, tableaux, gipsies. Albinoes, fat boys, giants, dwarfs," and dozens of other assorted attractions.[33] In stark comparison, the Smithsonian, the Columbian Exposition, and other cultural institutions had begun to rein in such freewheeling displays in favor of neatly ordered exhibits that emphasized completeness and taxonomy over variety. The *Progress* was a decidedly *anti*-modern rendering of museum craft from this vantage point—more reliant on the legacy of early showmanship than a newly-rising passion for exhaustive cataloging and order.

And there was one more, perhaps surprising person who knew that a more faithful museum to whaling was not only possible, but had been explicitly discarded in favor of this embrace of spectacle and marine fantasias: Captain Daniel W. Gifford. That Captain Gifford had been regularly annoyed and frustrated by his voyage experience was no secret. The press often delighted in casting him as another "salty tar" who wanted nothing to do with the throngs who approached his vessel or the freshwater routes he was captaining.

But once he returned to New Bedford, Gifford opened up about the museum he left behind in Chicago, and he was clearly disappointed, starting with that ill-fated discharge of the crew in Racine. Although Captain Gifford had carried out his duties and the letter of the contract the crew had signed, he made the point back in New Bedford that the "foremast hands have all been discharged," making sure his local brethren knew there was no crew of New Bedford whalers left back in Chicago. He also predicted that Captain Dowden himself would be returning soon because, "the Progress is not to be exhibited as a whaler."[34] He was perhaps responding to a local article that as late as August 1892 was still insisting to New Bedford audiences that "the purpose of her presence here is to give the people of Chicago and her visitors correct exemplification of the whaling industry."[35] If there were no New Bedford whalemen left on the *Progress*, how was a "correct exemplification of the whaling industry" possible? Captain Gifford wished to set the record straight.

Instead of an authentic depiction of whaling, Gifford saw that the old whaleship had been fitted with electric lights, had "lots of clerks on board, including women," and that the entire enterprise had been transformed into something "devoted to the sale of shells." On this last point he was not entirely wrong. The *Progress* did seem to have some sort of "exit through the giftshop" structure, with purchase options that included souvenir tiles from

Nine. Chicago, 1892

New Bedford's Bliss & Nye, purveyors of fine house furnishing goods.[36] Captain Gifford then went further, claiming that there were those who wanted to follow through on the original idea of a fully outfitted whaling ship that could demonstrate the trade accurately to fair audiences. "Capt. Gifford says there are parties in Chicago who wish to purchase a whaling vessel for exhibition as such, and he is now looking for one."[37] The fact that Gifford even articulated this idea suggests it was clear to him and others that (1) the original vision of an authentic whaling museum had collapsed and (2) that alternative versions of the *Progress* could, and in his estimation *should*, exist.

Gifford's condemnation of the enterprise in the press also had the unexpected side-effect of allowing Fitch Crane one more round of publicity, albeit through one of the strangest stories yet associated with the *Progress*. After spotting Crane in New Bedford in September 1892, some reporters wondered if Captain Gifford's prognostications of a second whaleship at the fair might be true. Crane seemed happy to feed the speculation, and even claimed to have purchased a bark in rough shape called the *Palmetto*. Crane went further still, promising to "fit her out at once to take her around to Chicago to exhibit her as a whaler and will purchase articles used in the whaling business."[38] The problem with the story is that it was entirely false. The *Palmetto* was never sold, her *Whalemen's Shipping List* record never changed, and she remained in New Bedford in state of watery decay until she was finally condemned in 1895, never coming close to seeing Chicago. But Crane's story of single-handedly purchasing a whaleship (on a livery supervisor's salary apparently) was picked up by the press, and for a few days at least Fitch Crane's name was out from under the shadow of Henry Weaver and back in the newspapers.[39]

By the end of the month Captain Gifford would be mastering another whaling voyage aboard the bark *Gay Head*. His instructions were to return the vessel at the end of his voyage to San Francisco instead of New Bedford—another ship lost to the West Coast. He was back among a real crew, on a real whaling voyage, with not a single string of electric lights or women selling souvenirs in sight. But his labors this voyage had ramifications far beyond just the pursuit of whale oil and whalebone. Gifford had also been tasked to use the voyage to travel to Antarctica, a destination that had not been whaled since 1871. Captain Gifford's instructions were to "see if a new hunting ground cannot be opened to commerce."[40] He and the *Gay Head* were due back in San Francisco around October 1894.

The contrast in the roles of these captains, Gifford and Dowden—who had shared the deck of the *Progress* just a few weeks earlier and had both been given "three cheers" by a gathering of Chicago elite—could not be more striking. Dowden had become a conduit to the past, tasked with being the human counterpart of the "marine curiosities" that filled the decks of the whaleship. His job was to help convert whaling from a violent, complex, and gory trade into something safe, amusing, fun, and consistent with a benignly hazy slice of American memory. Gifford, on the other hand, was still engaged in the actual trade of whaling and was tasked with trying to save the shrinking industry. He was a reminder that despite all the epitaphs and eulogies for whaling, the trade continued on in significant ways, particularly to those who still fought to keep it alive. "Next to nothing is known regarding the Antarctic Ocean," wrote the *New York Times* of this particular voyage that departed from New Bedford on August 29, 1892, with Gifford in command, "and the hope is entertained that the present expedition will develop a new field to replace the old one that is dying out."[41]

Still, as news of the *Progress*'s shift from a museum in which whaling was "so worthily represented," to a museum of marine curiosities filtered back to New Bedford, it became obvious that the entire enterprise was already being forgotten and ignored. Sad as the failure to represent whaling might be, it faced difficulties in making an impression on a community that had already said their goodbyes. When one of the city's socialites returned from a preview visit of the Exhibition in February 1893 she described everything she saw including Massachusetts' state building, along with halls dedicated to Agriculture, Transportation, Machinery, and the Women's Building. She made no mention that New Bedford's own *Progress* was also there.[42]

Attentions were elsewhere too as New Bedford experienced yet another unprecedented boom in cotton mills. Three new cotton mills opened their doors by August 1892—the Pierce Manufacturing Corporation and Rotch Mills in the spring, and The Columbia Spinning Company (a nod, perhaps, to the World's Fair excitement that had bubbled in the city earlier that year), in August. Writing his history of New Bedford that year, Leonard Bolles Ellis estimated that more than 12,500 people were now employed by the city's cotton manufacturing industry. By contrast, he concluded of whaling: "It must be conceded that the prosecution of the whale fishery has ceased to be of great importance to the community, and there is not prospect for its future growth and development."[43]

The Rotch Spinning Company—as it was originally chartered—was a particularly poignant note in this evolution, given that the *Progress* had set sail from Rotch Wharf that June. The wharf had been chartered in 1831, over sixty years earlier, older even than the *Progress* herself. In the words of one historian, "The Rotches were to whaling what Andrew Carnegie and John D. Rockefeller were to steel and oil."[44] Now the future, and the Rotch name, lay in cotton manufacturing, a fact not lost on at least one jokester during the

Nine. Chicago, 1892

city's semi-centennial just a few years later who quipped: "The city's motto, which adorns the city seal, shows the tendency of the times. It is 'Lucem Duffundo' [sic]—freely translated, 'I shed light'—and in the foreground appears a whaling bark sailing up the harbor. Were a new city seal to be adopted today ... the whaleship in the foreground would give way to the representation of a cotton mill, and a new motto would read: 'First in fine cottons.'"[45]

All of this quietly percolated along through the summer of 1892, with Captain Dowden and the *Progress* perched off State Street, Captain Gifford out at sea headed towards uncharted whaling grounds, and New Bedford's captains of industry busily getting thousands of new spindles up and running. Little attention would have been paid to the *Progress* and the museum she had evolved into until it was time to move her to the fairgrounds in the late autumn. She would have played a minor role in the zeitgeist of a city's summertime amusements such as when one cheeky ad suggested: "A trip to this 'old toiler of the sea' is certainly most invigorating and enjoyable, but no more so than a refreshing drink of Hygeia Wild Cherry Phosphate."[46] Instead, a disaster struck that thrust the *Progress* back into the limelight and began a chain of events that would ultimately define the whaleship for the remainder of her life.

The accounts of what happened on Saturday, September 24, 1892, further reveal the kind of attraction the *Progress* had become. According to the *Tribune*: "'Supes' who learned the art of navigation from steering South Water street schooners, attired in jaunty spring-bottom pants and décolleté blue shirts were leaning against the bulwarks or sitting about on coils of rope, chewing hard tobacco and otherwise lending nautical color and charm to the *Progress*."[47] Here is the most complete description yet of how far the attraction had shifted from its original idea and purpose, confirming Captain Gifford's worst fears. Now the men aboard were being described completely as actors, drawn not from authentic whaling stock (who had all been discharged, save "Fiji King" Jimmy Kanaka), but local boys who sailed schooners through freshwater routes. The attention to their costumes completes the transformation of the crew from real world workers to those playing a part and adding ambiance, and little else, to the visitor experience.

Again, it is impossible not to consider Steele MacKaye's influence on Henry Weaver at this juncture. By September, Weaver had officially bought into MacKaye's Columbian Celebration Company, was serving as a director of the company, and had publicly endorsed the Spectatorium project in the press. MacKaye would have deeply admired the fact that the *Progress*, and her bulwarks and coils of rope, could serve as authentic stage props for the actors on deck. It was exactly what he was seeking for his own Christopher

The Last Voyage of the Whaling Bark *Progress*

Columbus production. "The desire for authenticity dictated much of what was to be included," writes one historian of the Spectatorium, "On this account, MacKaye sent researchers to Spain to gather information on authentic costumes, dances, and architecture.... If props could not be adequately constructed (palm trees, for instance), the actual objects were to be used, regardless of expense."[48] Attention to details like the "jaunty" costumes the sailors wore, and the fact that Weaver was comfortable replacing real whalers with local substitutes, further suggests this attraction was now much more akin to a MacKaye production than the "correct exemplification of the whaling industry," that had still been described just a couple months earlier.[49]

Gone too, was Captain Dowden, who had returned to New Bedford a week earlier, and instead readers were introduced to Levin Olsen (or, in another account, Albert), "the real sailor of the ship," along with the ship's manager.[50] Also on board by around 2:00 p.m. were approximately 200 visitors, most of them children on account of Saturday being "Children's Day" for the museum. It was an impressive number of visitors, and according to the *Tribune*, the *Progress* had consistently been "a good drawing card."[51] Traffic around the State Street Bridge would have been heavy on a Saturday. The truss swing bridge would have regularly been rotated throughout the day to make room for two channels of traffic up and down the river. When it swung open that early afternoon to allow the tug *James Hay* and a heavily-laden sand scow to pass on their way upriver, few would have noticed.

Unfortunately, a large freighter headed downstream towards the lake was also passing through the swung-open bridge at the same time. The current and wake off its propellers caught the scow, and sent it sheering off from behind the tow and across the river. Its momentum across the river only stopped when it hit something big enough to stop it. Unfortunately, that something was the wooden hull of the *Progress*. The scow was all sharp corners; the *Progress* was a solid sheet of oak. The two collided, and in words of the *Tribune*, the cracking of wood "could have been heard a block away."[52] Most people aboard the *Progress* were likely holding their 4-page brochures, which ironically announced on page one that the ship was built "...of the best oak, with huge timbers and sides that from their thickness would seem to defy any power to crush or injure them."[53] That challenge was met and answered, and not in the way the brochure had predicted.

Estimates varied on how large a hole the scow created. Initial accounts put it from "two feet square" to "eighteen inches long to six inches wide," while the final assessment "found that a hole six feet long and three feet wide had been made."[54] Two things happened simultaneously—water immediately began rushing in, and the stairway between the wharf and the *Progress* cracked apart. Everyone on board was now effectively marooned on a ship that had a hole the size of a large man in its side. Water from the Chicago River was cascading through at an alarming rate. As men, women, and children were hurried up from below deck, Olsen and Foote tried to figure out how to solve the exodus problem. A sturdy plank was found that did the job of re-bridging the gap between ship and shore, but everyone would have to exit single file.

Accounts of what happened next took an almost humorous turn. Nearly every article commended the multitude of children, suggesting that their school training had made them perfectly calm and compliant. According to ship's manager Foote, the children "showed to advantage the benefits of the panic training given in the public schools," while another retelling suggested "the training the children had received at schools to prepare them for fires" had turned them into a well-regimented corps for the impending exodus.[55] Other accounts also gave credit to the sailors, who "were stout of lungs and arm."[56] The one group who did not receive accolades were the adult men. "Men fought him [Olsen] and other members of the crew to get off ahead of the children, but Olsen, Foote, and employees held them at bay," wrote the *Tribune*, with the *Inter Ocean* adding, "In fact, the people who were most excited were the men, and they were the first to rush ashore."[57]

Meanwhile, there was no doubt the *Progress* was sinking and sinking fast. "The ship filled at once and listed over to the port side.... The last of the party had barely reached the dock when the old ship, which had been gradually filling with water, settled down to the river bottom, leaving only the upper deck bare." One report put the elapsed time at 20 minutes, another at 10 minutes, while the official court testimony from the ensuing lawsuit suggested it only took an astonishing eight minutes for her to go "to the bottom of said Chicago River" where she was "submerged under the water thereof above its upper deck."[58] Other details from the various accounts highlighted the drama of the whaleship listing away from the wharf: "The distance between ship and dock was increasing as the ship filled, but every one [sic] was removed safely before the ship settled. There was only an inch of each end of the plank resting on supports when the last person got ashore."[59]

Once the human life was accounted for, reports turned to the artifacts aboard. Most acknowledged that some of the most valuable pieces were saved. "All the furs, relics of the Greely expedition and most of the very valuable articles were taken out and placed in express wagons," reporting the *Evening Post*.[60] The *Inter Ocean* credited the last person on shore as "Miss Emma Emerick, who had charge of the curios," perhaps one of the female "clerks" that Gifford had groused about.[61] "Some of the curios will be damaged," wrote the *Tribune*, "although some of the more valuable ones were either removed or in the weather side of the hold, above water."[62]

But clearly, much damage to the museum and her artifacts had been done. The *Inter Ocean* painted the bleakest picture: "Handsome cases filled with valuable exhibits from the Arctic regions float about the cabin and main deck. The upper deck is strewn with the skeletons of huge whales, tusks, jaws, shells, boxes of coral and other curios. The Arctic furs and Esquimaux clothing stowed away below will be badly damaged."[63] Outside of the public eye, the court case later that year painted a similarly grim picture: "…said Chicago River is used for and as an open sewer throughout its entire length and the water thereof which filled the hold of said ship 'Progress' in consequence of said collision defiled every thing it submerged."[64]

In true Gilded Age Chicago style, the various news outlets of the day also began calculating the monetary and liability implications. "Although the Progress was lying almost in the draw of the State street bridge," reported (nearly verbatim) both the *Tribune* and *Inter Ocean*, "the liability of the tug Hay for damage done by her tow is not doubted by marine men, and the tug will most likely be compelled to pay the entire loss, including the wrecking expenses." Both accounts also added, "This will amount to nearly as much as she is worth, but by marine law no more can be collected than the value of the tug and scow."[65] The *Evening Post* also reassured readers that salvaging the *Progress* was possible, "All that we will have to do is send a diver down, repair the break, put in the pumps, pump her out, take down her spars and she will be as good as new."[66]

The subtle reference to the *Progress* "lying almost in the draw of the State street bridge" did not go unnoticed by the *Milwaukee Sentinel*. "The expressions of satisfaction that were made by vessel men when they heard of the sinking of the boat led to the report that the collision was intentional. The Progress has been in the way of tugmen ever since she was tied up at the State street bridge."[67] Chicago's *Journal* laid the problem out in even more detail, adding that "many efforts have been made to secure her removal. The tugmen alleged that she blocked the river in its most crowded and dangerous part and that it was an imposition to allow her to remain there. Their protests were ignored, however, and boasts have been made that the ship would be run into some night and sunk if it was not taken away."[68] It is easy to assume that Weaver had used his political connections to swat away these complaints in order to keep his attraction in a prime location. Now that decision had nearly destroyed the very attraction he had been trying to promote.

The ensuing court case also helps flesh out a picture of what happened that day. While Chicago papers were confident in Henry Weaver's case against the tow company, the latter made a few pointed critiques of their own once they had the ear of an admiralty judge. First, the tow company argued, it was the freighter which had churned up the wake that really was the culprit. Not only did the *Fred Pabst*—named for the local brewer whose beer would soon win an eponymous blue ribbon at the Columbian Exposition—set off the series of events but it actively obstructed the tow when it tried to bring the sand scow under control. But more importantly, the *Progress* was not a "tight, staunch, and strong" ship as Henry Weaver claimed, but was in truth "so old, rotten, decayed and weak that she was unseaworthy and not fit to lie at a dock in the Chicago River where vessels were constantly passing and were liable to touch her in passing." Finally, the tow company reiterated the complaint about the *Progress* being there in the first place: "it is very difficult to navigate the said draw unless it is entirely unobstructed, and that said 'Progress' was, therefore, laying where she had no right to be and where she absolutely obstructed navigation through the draw of said bridge for all but small vessels...."[69]

All of this, of course, was riddled with symbolism—some explicitly recognized in press accounts, some not. The *Tribune* stated it most obviously: "For nearly half a century the gallant whaling vessel Progress braved all the countless dangers of the dark, damp, and fathomless deep. It was buffeted by the mighty waves of the all the seas. It came unharmed from out of the awful grip of icebergs.... It grew proud of its ability to keep out of trouble after it had visited half the ports in the world, and got around the Horn a score of times. The Progress is now resting peacefully at the bottom of the Chicago River."[70] The ship that had braved countless dangers across the seven seas couldn't survive two months in the heart of Chicago.

Other references were more opaque. Certainly the suggestion that sailors rejoiced at her sinking because she was "in the way" was reminiscent of the tugboat captain who thought the "old tars" of the *Progress* moved too slowly and wasted his time on the waters of the Great Lakes. The idea of an ancient vessel literally standing in the way of Chicago's modern commerce, navigation, and transportation clearly struck a chord with those who thought (some gleefully) that the sinking might have been intentional. Others suggested that the dramatic sinking made a far better spectacle than the floating museum herself: "The Progress never was half the curiosity in [the] Chicago River she was yesterday afternoon and last night. She went down with her flag nailed to the mast, so to speak."[71] This sort of swipe had also been made before, as when it was suggested that few would be interested "the museum of whaling appliances she may have on board." Across all these observations lay a disturbing thought—that the old time whaling ship might have been better off quietly decaying, like the whaling industry itself, back in "the mackeral-scented precincts of New Bedford."[72] The assertation that she was "not fit to lie at a dock in the Chicago River," had been entered into official court testimony. Was it true? Perhaps so. Maybe here in Chicago the museum was out of place, out of time, and ultimately, unwanted.

If Weaver read any of the press' slings and arrows he ignored them and went to work getting the *Progress* back above water. Divers were already at work the next day, Sunday, September 25, patching up the hole early that morning with the *Progress* tilted above their heads at a 45-degree angle. "The break was in such a position, too, that it could not be mended by an ordinary patch, but the planking had to be built out to a corner in order to make the turn." Just as had been described to readers the day before, once the patch was completed, pumping began and "the Progress slowly came to the surface, righting herself on the keel as the water was pumped out."[73] It took until around midnight before the pumping could be declared complete. On Monday she was put into drydock.

Unfortunately, by then the damage to artifacts was clear, including an unexpected problem—theft. Apparently, thieves had come in the cover of darkness on Saturday night "and carried away what they could lay their hands on."[74] What was not stolen was caked in mud, or thinking of Chicago's sanitation system in the 1890s, much more vile substances. "After the water was pumped out between the spar and main decks the hundreds of curiosities from the Arctic regions made a grotesque appearance under the light of the lanterns," reported the *Tribune*. "The filth of the Chicago River was over everything. The fancy work of the South Sea Islanders and the full dress suits of the Esquimaux looked so nearly alike that they could not be told apart. It will take many days' work to remove the sewage, and some of the articles are totally ruined."[75]

By now the story was also gaining national attention. A short, syndicated description spread over the next two days from New York to Indianapolis and Rock Island, Illinois, to Los Angeles and nearly every point in between. They all described the sunken *Progress* as a "museum of curiosities" that would be raised and repaired.[76] Meanwhile, Chicago papers assured readers that "Mr. Weaver, President of the company owning the Progress, will take steps today to fix the legal responsibility of the loss" and that $10,000 of marine insurance was carried on the *Progress* so that "the only loss will come in the cessation of the exhibition business and some of the curiosities which cannot be replaced."[77]

It would take until mid–October before the *Progress* could return to display at the State Street Bridge. It would be a limited-time engagement, as she

Nine. Chicago, 1892

was due to be towed to her spot at the fairgrounds shortly. She needed to be placed in the South Pond before it was closed off by the pedestrian bridges that would be built afterwards. The October 15 *Tribune* did what it could to send visitors back to the attraction with the headline: "Again at State Street Bridge—Great Museum Aboard," encouraging visitors to "spend an hour or two in her museum" now that she was out of dry-dock. It would be a little over a month before the *Progress* went off display yet again when she was towed to the fairgrounds in late November of that year. It is unclear how many visitors were willing to pay to come aboard and risk another mishap during this second downtown engagement.

Within that October 15, 1892, *Tribune* piece there was also embedded some new information: "The museum is to be presented to the Chicago University after the close of the Fair, and is one of the rarest of its kind in this country."[78] This news presented a brand-new facet of the *Progress*'s evolution, proof that no museum is ever static. While the idea of presenting the cases of objects to Chicago University was hardly a return to the idea of an authentic and representative depiction of whaling, it was an acknowledgment that the collection might have value beyond the quaint, curious, and attractive. Objects worthy of a university's interest elevated the prestige of the collection beyond the props and stagecraft that had dominated for much of the summer and early fall before the accident. What the article did not specify is what constituted the "rarest" quality of the collection. Was it the rarest collection of whaling objects? Arctic specimens? Marine animals? Objects of foreign cultures? The reader was left to wonder … and wait.

Which is exactly what happened to the *Progress*. Once towed to the fairgrounds on November 26, 1892, the whaleship went off display. Construction crews would erect all manner of bridges, sidewalks, and buildings around her, locking her into place on the South Pond while a White City sprang up on all sides. No ticket buyer would walk onto her deck again until the fair opened on May 1, 1893. During that time, the whaleship was not visited, was rarely observed, and was largely forgotten. There were a couple of wintertime press references that remarked how the whaler sitting at anchor was encased in ice, reminiscent of her Arctic adventures: "Snow rests on the spars and icicles hang from the cross-yards. The whaling boats are swung aboard at the stern just as they are when everything is made snug on board a vessel during a sojourn within the Arctic circle."[79] When builders could brave the cold or could scramble in the thaw of spring to catch up on their work, the contours of the *Progress*'s little corner of the fair took shape. Off her starboard side emerged the long, windowed façade of the shoe and leather building as well as a segment of the elevated rail track. Tucked just at the corner by shoes and leather

The Last Voyage of the Whaling Bark *Progress*

was the dairy building, while forward on the starboard side sat the turreted, castle-like structure commissioned by the Krupp gun company. Behind her were buildings dedicated to the French colonies, including a Tunis Cafe, while across the pond on the port side, a display of windmills would eventually fill the muddy ground. As all of these structures eked into existence during the months before the opening, the *Progress* sat as a silent sentinel.

It is a curiosity to wonder how far in advance Henry Weaver realized his attraction would be shut down for half a year, generating no income, while it sat waiting for the fair to be built up around it. Did Weaver know this would happen when he bought the ship in New Bedford? Was it explained to him later? The record is silent. What is certain, however, is that it was also a serious blow to the excitement and "buzz" that the *Progress* had generated for

Progress at anchor on the South Pond, Columbian Exposition, 1893. This photograph may have been taken as the fair's construction was nearing completion, just prior to opening in May 1893. Although the Krupps castle and the Shoe and Leather Building look completed, the grounds are still relatively bare. The photograph was originally published in an unknown German publication. The article was likely describing the proudly Germanic Krupp Gun Pavilion just off the *Progress*'s starboard side (courtesy Field Museum, A101644).

Nine. Chicago, 1892

much of 1892. The sinking into the Chicago River and her subsequent solitary confinement in the South Pond seemed to take all the wind out of her sails, both literally and figuratively. Weaver pursued his damages lawsuit against the towing company that had been pulling the sand scow in September. He also commissioned a 46-page souvenir book entitled *"There She Blows" or The Story of the Progress*. In an ideal world, the book could have served as another round of publicity for Weaver's venture, but with no ship to visit there was no logical place to sell it.

Although visitors would have to wait until May to purchase the book (another item for sale aboard the *Progress*, along with the aforementioned seashells and decorative Bliss & Nye tiles), *"There She Blows"* is still a fascinating window into Weaver's mind during these lost months of publicity and revenue. The author that Weaver secured was a perfect match to the project. G.A. Coffin grew up in Fairhaven, Massachusetts, and was himself a New England boatsman until his move to Chicago. There he opened a freelance business, making drawings for engravers, and producing commercial lithographic designs and paintings. As he developed a reputation and client base he was increasingly able to specialize in maritime subjects, and he became a regular contributor to the *Tribune*, which had frequent need for marine drawings. In fact, the *Tribune*'s Sunday feature on the *Progress* prior to her arrival in Chicago had been illustrated by Coffin.[80]

At Weaver's behest, Coffin set about writing a short fictional book for young readers about the whaleship. The story centers on two main characters from New Bedford—Jack, an old-time whaler, and Frank, a local boy whose uncle is one of the *Progress*'s New Bedford owners. Coffin quickly describes the two in a way that assures the reader of their goodness and moral fitness. Jack was "a not uncommon type of the old whaler ... small eyes that twinkled as brightly as those of St. Nick, a frank open face, which all the seams and scars of a half century's battles with the elements could not change, gave an air of rugged honesty and sincerity, that never failed to win the heart of his youthful admirers, or the confidence of his employers." Frank's treatment is similarly inviting and modellable for young readers, "as bright and manly a lad for his age as you could find."[81] The story begins with Frank and Jack sitting aboard the *Progress*, and the lad telling Jack that he has learned his uncle will be selling the *Progress* to some Chicago men for display at the fair. As the hearsay becomes reality, the two decide to journey with the *Progress* to Chicago, so that along the way Jack can tell his stories about the whaleship and whaling days to his young listener.

The book intersperses the events of 1892—the sale of the *Progress*, the rush to prepare her for departure, the journey to and display in Chicago (although it studiously avoids any mention of the sinking into the Chicago River)—with Jack's stories going back to the ship's construction and original launch as the *Charles Phelps*. Daring whale captures, destructive gales, the rescue of natives, and of course, the Arctic disaster of 1871 are all told in the typical sensational style of the period's dime novels as well as serialized stories in youth-directed periodicals. Coffin's illustrations and a few photographs are sprinkled liberally throughout.

Produced specifically for the visitor to the World's Fair, the last paragraphs bring the reader up to the moment that they themselves enter the story by boarding the *Progress*: "And now my readers this story is told. The ship is before you. Use your eyes and you will not be disappointed." Given the number of references to storybooks and dime novels the *Progress* encountered prior to the fair, it is a fascinating rhetorical device. "Old Jack, the leading character of this story, is with the ship today," readers are informed, although who would be playing the part of Jack was a bit of mystery.[82]

There was also the fact that readers took the journey of the *Progress* from New Bedford to Chicago with Jack and Frank. Proportionally the retelling of these events makes up about one-third of the book compared to the two-thirds made up of historical whaling stories. Still, it is a significant commitment on the part of Weaver (through Coffin's work) to insert himself and his purchase into the narrative. It reveals Weaver's assumptions about the *Progress*—that not only would visitors be awed by the ship, the museum, and the ship's salty tars, but by Weaver himself and the "pluck" (in the words of Benjamin Butterworth) it took to bring the whaleship to Chicago. Weaver is written into the story as an unnamed Chicagoan whose very presence "spoke of large sums of money, and big enterprises in such a matter of fact way as to almost take the breath away from some of his hearers." Lest someone not recognize this "successful Chicago businessman" as Henry Weaver, he also inserted a full-page ad for his coal company at the end of *"There She Blows."*[83]

This editorial decision to inform readers of the sale and journey was not solely an act of hubris or vanity on the part of Weaver, although it certainly was that. It also revealed the continuing assumptions being made about the *Progress*—that everything about her and her story would be of interest, note, and importance to Americans in the 1890s. Just as New Bedforders had made assumptions about the draw of an authentic whaling museum, so too did Weaver make assumptions about how his own role in the story would be seen. Just as New Bedforders had come to believe their own evangelical zeal for whaling, so too had Weaver come to believe the flattery of his peers, suggesting that he had done something unique and important in bringing the *Progress* to Chicago. Of course, both sets of assumptions remained untested

Nine. Chicago, 1892

by the fair's visiting public until the grounds were eventually, finally, thrown opened to its millions of visitors. Maybe these assumptions about the entire enterprise would turn out to be accurate; maybe they were just pipedreams. But for now, the *Progress* sat, waiting for the fair to commence and for visitors to resume their acquaintance with her museum of marine curiosities, and to help settle these questions in the final analysis.

They did, but not in the ways anyone involved would have hoped. By all accounts, visitation to the bark was anemic throughout the fair's run from May to October 1893. In short, the *Progress* was a bust. Part of the problem was likely the admission charge. The 25-cent entrance fee was about the equivalent of a full-price movie ticket today. So, while it was not an out-of-reach expense for most people, it was also not a casual one.

The *Progress* was not part of the Midway—the raucous, freewheeling set of attractions that ranged from camel rides to the famous Ferris Wheel to a recreation of Ireland's Donegal Castle, all of which charged admission fees. The *Progress* was situated within the "White City," the venue of neoclassical palaces dedicated to art, industry, and the new world order of the Gilded Age. Every corner of the White City was free with one's overall admission ticket save for three things: the electric boats that plied the lagoons and North Pond; the "Cliff Dwellers" exhibit, which recreated a one-sixth version of Anasazi ruins found in Colorado; and the *Progress*. This left the floating museum as something of an outlier. Although some fair reports promised that the *Progress* "contained a marine museum of considerable interest," other official materials had suggested that only "two or three minor concessions" would be charging fees—hardly a ringing endorsement.[84] The bottom line: the added charge was likely to cause visitors considerable pause before parting with 25 more cents per person.

The *Progress* was also now surrounded by competition. When displayed in places like Montreal, Buffalo, or Racine, or while anchored off the State Street Bridge, the admission price of a quarter seemed perfectly normal as a standalone attraction. The problem now was that it was not a standalone attraction. Now it was one of thousands—tens of thousands—of choices that a visitor had to navigate when visiting the fair.

Within the myriad of guidebooks published for the fair's multitudes (it is estimated a quarter of the entire U.S. population visited the fair) there was little that made the *Progress* stand out. *Conkley's Complete Guide* delivered less than 25 words in its "Points of Interest" section: "Whaling Bark 'Progress,' South Pond—the old whaling bark 'Progress.' A museum illustrating the whaling industry. Containing marine curiosities and relics. Admission 25 cents." The listing fell between the Vienna Maennerchor Society and the Electric Scenic Theatre.[85]

The Last Voyage of the Whaling Bark *Progress*

The attraction fared slightly better in Rand McNally's *A Week at the Fair*, earning a whole paragraph which promised something much more akin to the original idea for a whaling museum: "In its saloon are shown the articles usually obtained by or used in the whaling industry, as polar bear-skins, seal-skins, blubber, whalebone, knives, harpoons, tackle, boats, etc." The writeup also noted that the whaleship was "close to the Ethnographical exhibit" which had also sprung up along the shores of the South Pond.[86]

The fair guide from the Harper's company provided 90 words on the Progress and suggested, "She is a splendid example of the type of vessels which formed our once formidable whaling fleet of romantic memory."[87] Perhaps one of the best writeups came from White and Igleheart's handbook. The authors in this case actually attempted to place the reader on the deck of the *Progress*, "a veritable old 'blubber hunter,'" adding that the whaleship was "one of the most instructive object lessons at the Fair." The authors credited "her full lines, boats on cranes, try-works and general outfit" with teaching visitors about the history of the whaling industry and its importance as "a proud chapter in our national history."[88]

The problem was that even this outstanding writeup was on page 229 of a 628-page tome, buried in the verbiage of the "Fisheries" chapter. What were the odds that a reader, visitor, or prospective visitor would not only focus on this recommendation, but find it more enticing than the myriad of other offerings? How likely was it that readers would be steered to the *Progress* by Rand McNally's promises of blubber and seal skins? Nor was the *Progress* exactly placed in a "must see" spot within the fairgrounds, surrounded as it was by shoes and leather, dairy displays, a Tunisian cafe, and whatever the anthropology experts had dreamed up including a recreated Native American village complete with totem poles. In truth the entire South Pond was often derided as "the frog pond" with the large selection of windmills filling out the other side. The latter inspired *Puck* magazine to joke—unsubtly referencing this forgotten corner of the fairgrounds under the headline "In the 'Back Yard'"—"'Why have they got all those windmills together for?'; 'Why, after the Fair, they've got to blow that 'ere whaler into the Lake.'"[89]

Returning to the guidebook writeups that at least offered a chance to steer visitors to the *Progress*'s ticket booth, it is also possible to see a tonal shift. Whereas upon leaving New Bedford—and especially upon arriving the Chicago—the *Progress* had transformed into a side show-esque spectacle of marine exotica and carefully staged recreations of dime novel characters, now in the White City it was trying to reassert its credentials as educational and uplifting, perhaps even patriotic, as was apparent in the suggestion of "a proud chapter in our national history." Her place in the White City—instead

Nine. Chicago, 1892

Another view of the *Progress* (far right) in the South Pond, surrounded by Native American exhibits and across from the windmill display. From Hubert Howe Bancroft, *Book of the Fair*, Fin de Siècle Edition, Section Three (Chicago: The Bancroft Company, 1893).

of the more carnival-like Midway section of the fair, with beer gardens, dancing girls and thrill rides—may have been part of this metamorphosis back into a more educational and realistic paean to the whaling industry. Gone were gimmicks like Jimmy Kanaka the Fiji king.

The problem with this was two-fold. First, the renaissance was too little, too late. Without a prime position such as the State Street Bridge—as well as monopoly on such a unique offering—the *Progress* simply could not garner enough attention to secure a steady flow of visitors. Second, and perhaps more importantly, no one was interested in paying extra for a didactic display to a dying industry. One particular article wrote of the problem with scathing, unsentimental bluntness: "Some way or other fair visitors did not want to know how a whale was harpooned or how sailors tunneled into him and excavated his bones, or how his oil was boiled. The apathy which people evinced toward the whaling science was remarkable. Only a few visitors came and dropped a few miserable dollars into the large, roomy coffers."[90]

One of the most complete descriptions of visiting the *Progress* turns up in a children's publication: *The Century World's Fair Book for Boys and Girls*. Clocking in at nearly 250 pages, the *Century World's Fair Book* is an impressive tome dedicated to the fair experiences of young boys Harry, Philip, and their tutor. It is Philip who visits the *Progress* and over three pages describes it to his friend. The conversation starts with a notable jab to the *Progress*'s lackluster attendance: "Well, at first I was n't going in," Philip reported back to Harry, "for they charged a quarter, and there did n't seem to be many going on board. I was afraid it was not good for anything, but at last I made up my mind to risk twenty-five cents on it. I bought my ticket and climbed the gangplank. There were just two other men on board besides the sailor in charge."

Philip then describes what a visitor saw, heard, and experienced once on board. "The sailor came forward to speak his little piece. He said if we wanted to know how they caught whales he'd tell us. Then he went on with the whole thing, from 'Thar she blows!' down to cutting up and trying-out of blubber. I had often read about it, but I tell you, Harry, it was different to see him hold up the harpoon and the lance, the gun for firing a big harpoon and all." In fact, the *Century World's Fair Book* seems to describe what many in New Bedford had hoped for all along—a way to communicate the process of whaling to those who knew nothing of the trade and craft beyond storybook adventures. Philip's sailor/educator also introduced the boy to the tryworks and explained the value of whalebone versus whale oil.

"After he finished telling about whaling, he invited us below to see a collection of marine curiosities they had on board," Philip continued. "'It was a regular old-style ship, with beams coming close down to your head. All around were cases of curious things—real sailor's oddities.'" Here, then, was the shift from the didactic lessons on whaling to the marine curiosities, including "carved teeth and shells, swords from sword-fish, idols, weapons, tools." It is both a physical and mental transition, moving from the open deck's discussion and demonstration of whaling to a "space below," filled with "real sailors' oddities." Philip also describes how he bought a book about the whaleship—undoubtedly the very same *"There She Blows"* that Weaver had commissioned G.A. Coffin to write. And this allows him to make the third narrative link to the *Progress*, reminding Harry that the ship saved "the poor fellows from the ice. That was what I call square."

Thus, in *The Century World's Fair Book for Boys and Girls* we see how it was possible for Weaver and the Chicago owners to try and straddle all the versions of the *Progress* that had been conceived and attempted. Philip's description, while varied, is easy to follow and imagine—a whaling display followed by marine curiosities, with additional information about the Arctic narrative available for sale. But *The Century World's Fair Book for Boys and*

Nine. Chicago, 1892

Girls had one more commentary to make about the whaling display which cost fairgoers a quarter. For all his vivid retelling of the visit over three pages of the book, Philip's description had one effect on his friend Harry: "Philip turned to look at Harry more closely, and found that the tired boy had fallen fast asleep." There was perhaps no greater damnation of the *Progress*'s drawing power than the suggestion that describing an actual visit to a whaleship could put a young American boy to sleep.[91]

Despite author Tutor Jenks' helpful description, it seems that what was ultimately shown at the Columbian Exposition was a muddled mess—part original concept of a faithful whaling museum and part collection of curios and whatnots. When Mary Sherman reported from the fair for the *Los Angeles Times*, her write-up of the *Progress* revolved solely around the Arctic rescue story and the "museum of Arctic curios" in the decks, including "the fourteen hundred-pound turtle, the skeleton of a sea serpent, and the fur suit worn by Greely during his expedition." As was so often the case, Sherman also focused on the Arctic rescue with its sacrifice of cargo and saving of "260 survivors, among whom were women and children."[92]

Author Mrs. Mark Stevens—who had clearly picked up one of the State Street Bridge brochures and plagiarized heavily from it—mentioned briefly that a "lecture was given in broken English by the first mate of the boat, a German" and that the heavily accented presentation included a display of whale baleen. By comparison she devotes three paragraphs in her write-up to the sea turtle, a sawfish skull, the mummified Australian boy, "delicate laces which looked like linen, which the women of the South Sea Islands make of the cactus plant," and the quadrant that supposedly came over on the *Mayflower*. Captain Dowden seems largely absent too, with Stevens mentioning in passing that "he still had charge of her at the Fair," but doing little to confirm meeting him in person. Both women described Dowden's gold watch or medal, with Stevens adding they were accompanied by "a touchingly grateful letter."[93]

The *Progress* did garner some positive reviews. In a letter home to his parents in Poultney, Vermont, Ned Clark wrote that the museum below deck was "a corker" and "on the whole, we voted unanimously that it was the best thing yet." Yet even Clark's letter revealed that it was impossible to escape the narrative of death that was necessarily part of an exhibit about whaling. "Gathered around are the harpoons, lances, bomb-lances, guns, lines, etc. used in murdering these poor devils of whales...." And the dichotomy between the faithful exhibit to whaling on deck where each item was "in its own little place," as the Vermont visitor described it, versus the "10,000 marine curiosities" below deck simply failed to stir the imagination of most fair visitors.[94] *The Boston Globe*—that grande dame of Massachusetts newspapers—actually had the audacity to write in June 1893, "It seems somewhat surprising that some Massachusetts man or organization did not exhibit some of the floating relics of whaling days at the World's fair," obviously ignoring or conveniently forgetting that New Bedford had done exactly that.[95]

In the midst of all the fair excitement, at a time when Henry Weaver should have been putting more and more effort into the *Progress*'s success, he was instead distracted by his own difficulties. His company Weaver, Getz, & Co. went into receivership on May 29, 1893, just a few weeks after the fair had opened. This sort of thing had become increasingly common in the wake of one of the nation's biggest financial upheavals in the first months of 1893. The depression that followed took down banks and railroads especially, and spiked unemployment. Weaver's excess of liabilities over assets put his banker (and co-owner of the *Progress*) William D. Preston in the uncomfortable position of declaring on the record whether his business partner was a good credit risk or not. He opted for yes. "We have always considered them good," suggested the Metropolitan National Bank executive of Weaver's company, and thus of Weaver himself.[96]

Weaver's financial situation was so bad that he even submitted a resignation letter to the Union League Club to relinquish his directorship—the three-year stint that he had just started in January. "My entire time will be required in the endeavor to straighten out my business affairs," Weaver wrote to his friends and peers, "and I think it only proper under the circumstances that my resignation herewith presented be promptly accepted." His friends and peers disagreed. It was moved "that the Secretary see Mr. Weaver and express to him the sympathy of the Board of Managers of the Union League Club for his financial misfortunes and request in the name of the Board that Mr. Weaver withdraw his resignation...." The motion carried, and Weaver remained a director, albeit one who attended far fewer meetings than he had in the past.[97]

With Weaver's focus clearly directed elsewhere, the *Progress* drifted further and further into obsolescence. None of the first-hand accounts cited suggested that Captain Dowden was still a central "character" or main feature of the exhibit. Whether he ever returned to *Progress* once the fair opened is unclear. Without him present it was only by quoting promotional materials that one was made aware of the 1871 drama of the Arctic disaster. Instead, the Janus-facing whaling exhibit above decks and marine curiosities below came to define the *Progress* to her few visitors. Neither seemed particularly compelling, Ned Clark's enthusiasm about "a corker" notwithstanding. The fictional Philip was not alone in noting that nobody seemed to be buying

Nine. Chicago, 1892

tickets to go aboard. One of her biggest nights came when the nearby Native American village on the South Pond put on a demonstration of war dances and costumes. Only then did a large number of people pay "to go on board the old craft to see better, and its deck, life-saving boat, and even the gangway was packed with people."[98] In fact, the reason *Los Angeles Times* reporter Mary Sherman made any mention of the *Progress* in her dispatch was because "mine was the privilege of viewing the scene from the deck of the Progress, an old whaling ship which was towed in before the bridges over the lagoon were built."[99]

The front entrance to the *Progress* or "Arctic Whaling Museum" at the Columbian Exposition. Photograph by John Reilly, Jr., 1893. This is an extremely rare amateur snapshot of the *Progress* at the fair. The photograph was taken by then–17-year-old John Reilly, the son of a U.S. Congressman and a blossoming photography enthusiast. Although most of his roll of film was dedicated to much grander parts of the Columbian Exposition, such as the Palace of Fine Arts and the Electricity Building, Reilly used one of his shots to record the entrance to the museum. A ticket booth can be seen at the entrance where the 25-cent admission would be collected. Visitors then walked under the banner promising "10,000 marine curiosities" as they boarded the ship (courtesy American Numismatic Society).

The Last Voyage of the Whaling Bark *Progress*

Perhaps the saddest day came on September 19, when the fair held "Fishermen's Day" to afford "all the fishing interests of the world a special opportunity to assemble in grand congress at the World's fair."[100] One of the highlights of the day was a procession of ships, a parade around the lagoons that would be led by a whaleship. However, hemmed in by the bridges that kept her confined to the South Pond, it wasn't the *Progress* at the front of this convoy. It was a "representation of a whaleship" dubbed the *F.D. Millet* to honor the fair's Director of Decoration, Functions, and Ceremonies, that was purported to be "a faithful representation of the old-time whaling ship." The fake whaler, helmed by an "old sailor" named Jimmie Hunt, who had been part of the fair's opening ceremony, would lead a parade that included a Turkish caïque, a fishing raft from Brazil, and a water bicycle. Only after all of these, third from the end before a "float with fishing camp" and a "sturgeon boat," would fairgoers see the *Progress*'s contribution to the great parade of boats—a kayak from her collection of Arctic curiosities.[101]

And then, it was over. The fair closed on October 30, 1893. Ironically, Henry Weaver's company had just come out of receivership a few days earlier.[102] By all accounts that would follow, the whaling museum had been a failure. But what, specifically, went wrong?

As had been pointed out all along the way, it was perhaps foolish to assume that people cared about an industry that was parochial, insulated, and complex. And perhaps it was foolish to assume people cared about an industry that almost everyone saw as obsolete and relegated to the past. The combination of the two was even deadlier. While true that whaling had romance and drama and a certain mystique wrapped up in boyhood tales of adventure and redemption, most people merely needed to gaze upon the old bark to stir up those feelings. Paying extra to try and recapture that spirit onboard simply did not make sense to the average fairgoer, particularly when the admission charge revealed a museum more eclectic than evocative. Others put a slightly different spin on it all: "The project of taking a New Bedford whaleship to Chicago to be exhibited as showing a representative world's industry, an industry which was then passing, however, started out as an idealistic venture. It was later turned into a commercial venture, however, a group of Chicago men purchasing the vessel to take her to the fair to be exhibited to those who cared to pay the entrance fee to see a whaleship."[103] Paraphrasing the article's thick prose: the commercial side of the venture failed to live up to the expectations and idealism that had launched it. A whaling ship from New Bedford would never "become famous not only at home but abroad," as those with a connection to the whaling industry had once hoped.[104]

The *Progress* was hardly the only failure of the fair. Steele MacKaye's vi-

Nine. Chicago, 1892

sion of the Spectatorium also never come to fruition. In the words of one scholar, "Ultimately, the project suffered because of the meddling investors. As a result of their inflated goals, construction was delayed and finally ground to a halt. It was never resumed when economic depression struck in the spring of 1893."[105] Henry Weaver, it must be remembered, was one of those investors. Both of his world's fair ambitions had now failed.

In the end, Weaver embraced MacKaye's theatricality and spectacle only to a point—perhaps most forcefully when the *Progress* was at the State Street Bridge. But when the *Progress* moved to the fair, these instincts retreated. The long winter and his own financial troubles had, of course, moved Weaver's attention elsewhere, and who is to say if continuing such theatrics would have changed anything. Ultimately, like the Spectatorium, the *Progress* suffered from too many ideas—whaling museum, curiosity cabinet, demonstration stage, ethnological collection, storybook set, patriotic memorial, heroic commemoration—none of which were guaranteed to substitute success for failure had they been the sole vision.

The only thing that is known for sure is that the end result was a far cry from what the citizens of New Bedford had hoped for back in May 1892 when they sold the *Progress* to a group of men from Chicago. Ironically, just a few days before the fair closed, New Bedford's *Republican Standard* published a piece about the city's collection of decaying whaleships. "Our older residents who were identified with the whale fishery in those halcyon days are pardonable if they invest these old hulks with a sentimentality that they do not possess for the younger portion of the population, and if they express deep felt regret at their disappearance."[106] And now—especially for those who "identified with the whale fishery"—things were about to get much, much worse.

Ten

New Bedford, 1904

William W. Crapo looked around the room with satisfaction. Less than a year ago the Old Dartmouth Historical Society (ODHS) had not even existed; now Crapo was its President. Not only was the organization a reality, but it had an energized Museum section which had already pulled together a loan exhibition within their leased rooms of New Bedford's Masonic Building. Crapo walked around his organization's first exhibition on opening night, February 15, 1904. He and the other guests were drinking punch and listening to music from the Gray & Sullivan's Orchestra. The entire affair was bedecked in red, white, and blue, a striking juxtaposition with the multitude of objects from the South Seas and Arctic regions. More importantly, the whaling artifacts were putting on a particularly good show, described as "the greatest collection of whaling relics in the world."[1] It was a great beginning, but Crapo was determined to build on the momentum for a permanent museum to whaling.[2]

Crapo had returned to New Bedford in the 1880s after nearly a decade serving as a U.S. Representative, and now practiced as a lawyer. He was also at the center of New Bedford's cotton manufacturing boom, serving as a director of both the Potomska Mills and the Wamsutta Mills where he was currently acting as President. By some accounts he was known as the "Dean of the Cotton Industry." His business interests spread into real estate, railroads, banking, and Republican politics, but these days, he was thinking more and more about whaling. Though never a whaler himself, Crapo had grown up surrounded by the industry and had always been enthralled by it. "As a boy he was mesmerized by the tales of neighbor Everlasting Bill Howland, an old sea captain," notes his biographers, and he had "absorbed a steady diet of stories of and from the sea."[3]

Crapo and other citizens of New Bedford had been energized just over a year earlier on the night of January 16, 1903, by a Unity Club lecture given by a reporter with the *Evening Standard*. The speaker may have been a journalist, but it was not his profession that fed the fires of his message, it was his name—Ellis Howland. Not unlike Crapo himself, Howland knew that his name had

Ten. New Bedford, 1904

at one time been synonymous with whaling. And now that connection was being forgotten and obviated with each passing year. Ellis Howland had suggested the remedy: "It is almost criminal that there is in this city no general museum of the whaling industry," he told his audience. "It would be compounding the crime to allow the void to remain much longer. There are in existence several greater or lesser collections of whaling gear and trophies of whaling voyages which might be gathered into one common repository and would make a credible museum. It is not too late to undertake the task."[4]

"Portrait of William W. Crapo." Jean Paul Selinger, 1908. Crapo was the first president of the Old Dartmouth Historical Society (courtesy the New Bedford Whaling Museum).

The rallying cry had resulted in the formation of the ODHS, and the call for a permanent museum had been regularly repeated ever since. His whirlwind Director of the Museum Section, Annie Seabury Wood, was likely already preparing her report for their first annual meeting next month in March. "The long deferred museum," she would tell her fellow ODHS members, "which all New Bedford knew should have been started years ago, was at last under way."[5]

Yet while Ellis Howland's January 1903 speech was verbose and thorough, not once did he mention that only a few months earlier, the final remnants of New Bedford's *previous* museum to whaling had been dynamited to a muddy grave. Nor did anyone seem to comment that the New Bedford-collected artifacts that had been displayed on the *Progress* were eerily similar to those that Annie Wood now oversaw. Flowing through the doors of 40 Masonic Building were objects that would have been perfectly at home in the previous incarnation of a whaling museum: "some old maps and account books came along, a musket used in the defense of Old Dartmouth, a water color by Benjamin Russell, a sampler, some shells and an American and an English flag.... Our collection of whaling implements began to grow, and then we felt that our raison d'être was established." Ellis Howland made no mention of the *Progress* when he suggested "a treasury of romance and adventure" which could tell the stories of "the pioneer invaders of the frozen mysteries of the polar seas." The closest anyone came that night to acknowledging the past was when George Stetson stood up and said "that he had been surprised, during his 30 years residence in the city, that the people had so allowed their opportunities to go to waste. It was almost a crime, he said, to allow the material which had been in the city to be scattered."[6]

But perhaps a bit of historical amnesia was understandable. The story of the *Progress* had not ended in 1893 but instead had lumbered on for years, blackening the eye of the whaling industry and New Bedford, before finally being put to rest around the time Ellis Howland was thinking about and researching New Bedford's whaling history. Though a remarkable coincidence of timing, no one was willing to acknowledge the overlap between the muddy, stench-filled end of one whaling museum and the inspirational speech that would launch another. Some things were just better left unsaid.

When the fair closed, the *Progress* was effectively abandoned onsite. Weaver walked away from the ship, essentially washing his hands of it. By now the suggestion that the original crew had 18-month contracts so they could bring the *Progress* home to New Bedford seemed positively laughable and quaint. Weaver did, however, make good on one of the claims made along the way—the collection of artifacts would find a new home. It was not Chicago University that received the materials, but the newly organized Field Columbian Museum. The Field was housed in the only permanent building that remained from the Exposition, the Palace of Fine Arts, so built because the priceless art displayed there during the Exposition needed a fireproof home for insurance purposes. Harlow Higinbotham had served as President of the Fair, and was now a trustee for the museum. He brokered the deal with Weaver, quite likely at the Union League Club where he and Weaver were both members. According to the *Tribune*, the collection was bought for $5,000; Weaver suggested the original cost was nearly $20,000.[7]

As the fair wound down and the financial wreckage of Weaver's boondoggle came into focus this would be one of many speculations into just how much money Weaver and the rest of the syndicate lost with the *Progress* venture. Another *Tribune* piece a few years later suggested that "When Henry E. Weaver turned it over to the Exhibit company it was shown the expense of bringing it to Chicago amounted to nearly $30,000, and at the close of the World's Fair the owners of the boat estimated their losses at $20,000."[8] Elsewhere the conjecture was that "altogether the vessel and its stock of curiosities cost its owners $40,000. From the result of the venture it seems they might almost as well have dumped the money in the sea for the enterprise proved a most disastrous one."[9] Still others were simply content to acknowledge, "The boat was a failure at the fair and the promoters of the scheme to bring it here abandoned the ship."[10]

Weaver had been cagey about what he had paid for the *Progress* from the beginning, and Ossian Guthrie had refused to tell Chicago reporters the sum when he returned from New Bedford. But the Board of Trade had already publicly laid out their plans back in January 1892, including the amount Bartlett wanted for the *Progress*: $3,200. It was consistent with what other whaling ships were fetching by that time. The J. & W.R. Wing company listed out several ships sold by 1891, averaging $3,300 per sale. Within a week of the sale of the *Progress*, Bartlett deposited a $3,000 check in his whaling firm's bank account. Although a copy of the check does not survive, the amount seems to exactly match what Weaver would have needed to pay to secure his whaleship.[11]

The problems with Weaver recouping his costs (of which $3,000 for the ship herself was just the beginning) stretched, in part, back to when the *Progress* was sunk in the Chicago River. At the end of the day, Weaver could not count on recouping losses from the damage done by the errant sand scow at the State Street Bridge. The initial ruling awarded him $4,576.77 in compensation. That was the good news. The bad news was that the case dragged on well into 1894 due to legal wrangling. In the course of the proceedings, the James Hay Company vessel was seized to help pay for the damages. In fact, the leaky scow was only valued at $1,200, well below what the ruling had awarded him. It is not clear if Weaver ever actually saw a dime in compensation.[12]

Compounding the difficulty with the guessing game about losses was the fact that the finances of the *Progress* had been strange at best and fraudulent at worst from the very beginning. When the Arctic Whaling Exhibit Company was organized back in June 1892 the original three investors—Henry Weaver, William Preston, and Edward Turner—were then authorized to sell $50,000 worth of stock in the company divvied up into $100 shares. By July 9 they filed with the Secretary of State's office that the stock was fully subscribed. Weaver and Preston had taken four shares each, Turner had two. Officers in the newly-subscribed company included E.W. Gillett from the Magic Yeast company at two shares, and Albert L. Tucker, a utilities executive, at one share. The rest of the roster included E. Fitch Crane (2 shares); J. Frank Aldrich, Chicago's Commissioner of Public Works (2 shares); Wardell Guthrie, another of Weaver's uncles by marriage who worked closely with Ossian (2 shares); and

The Last Voyage of the Whaling Bark *Progress*

George Bartlett, the original owner of the *Progress* back in New Bedford (1 share). In total, this amounted to 20 out of 500 shares—just four percent.

The other $48,000 was listed under the name Edward E. Carroll, a person who remains an absolute mystery with no other connection to the *Progress*, the fair, the Union League Club, or any known industry in Chicago. Despite owning 96 percent of the company, Carroll was not made a Director, did not participate in any of the *Progress*'s arrival activities in July 1892, and was never mentioned in the press.[13] Nor did Carroll's name come up when the *Progress*'s case from her sinking moved through both district and circuit court. In all of these court cases, news accounts, and event lists, it is solely Henry Weaver's name that is presented in connection with the Arctic Whaling Exhibit Company.[14] Given that $48,000 represents well over $1 million in today's dollars, it seems almost certain this was Henry Weaver's money placed under the name of someone else, but for what purpose? Perhaps Weaver was paying off a debt—there was a tradesman named Edward E. Carroll in Chicago who had formed the Complete Construction Company just the year before.[15] Whether the connection lay there or elsewhere, it was certainly an odd arrangement, one that clouded the finances of the *Progress* from the beginning.

The spotlight on Weaver's losses also challenged his own carefully crafted narrative of business acumen and prowess. This had already taken one hit with the placement of his company's assets into receivership throughout most of 1893. And although his friends at the Union League Club had overridden his resignation in a show of support, Weaver mostly steered clear of the Club and facing his peers during his so-called "financial misfortunes." The Spectatorium's failure had also been very public and was developing into a nasty legal contest between creditors and stockholders like Weaver. Even though he had come out of receivership, the *Progress* was still hanging like an albatross around his neck. The selloff of the collection to the Field Museum therefore worked double duty. Not only did it help put $5,000 back into the "black" side of the ledger, it also suggested that Weaver had done something that would contribute to the ongoing intellectual and moral uplift of the city through the objects he had assembled.

"The exhibit includes everything imaginable relative to the whaling industry," the *Tribune* reported somewhat ironically given that the focus on the whaling industry had been so frequently subsumed by the museum's "marine curiosities." Now they were being touted as a major selling point, as was the large scrimshaw collection, about which virtually nothing had been mentioned in nearly two years-worth of previous press accounts. "Among the rarest of the exhibits are several horns of the norwhale [sic]. Two of these horns of ivory are particularly perfect and are seven and one-half feet in length. Harvard College offered $500 each for these." Contrast this with the fact that throughout 1892 and 1893 the narwhal horns had barely earned a nod, and were sandwiched between starfish and "models of famous ships" in the State Street Bridge souvenir writeup.[16] Through this selective reimagining of the *Progress*'s collection, the Field acquisition allowed Weaver to save face from suggestions that he had overspent in assembling his collections. It was precisely because Weaver's collection had been so eclectic and far-ranging that the Field had the opportunity to report out to Chicagoans the objects they deemed of greatest value and significance.

When the museum opened its doors in June 1894 the published guide placed mention of the *Progress* at the very beginning of its "Section of Fishery Industries": "The Museum has been fortunate enough to acquire the very valuable collection which comprised the cabinet of curiosities of the Whale Ship 'Progress,' which, after many years' service in the Northern Seas, found a safe harbor in South Pond during the Exposition, where it attracted much attention." Forgiving the little white lie that it attracted much attention (or perhaps that it did attract attention, just not paying customers), the guidebook began in a familiar way, returning the reader to the notion that the collection was a "cabinet of curiosities."[17]

However, the museum guide then took an unfamiliar turn: "The decadence of the whale Fisheries during late years renders this collection a most important addition to the Museum. It is not improbable that the time may come during the life of the present generation when the sperm and the right whale on the high seas will be almost as much of a curiosity as the Buffalo upon the prairie." This was an entirely new argument. All the way back to New Bedford, the museum had been seen as a way of memorializing the dying whaling industry and accurately capturing the romance, drama, and history of a disappearing trade. Even as it drifted further and further towards an exhibit of "marine curiosities," those curiosities were still seen as collected by the hands of whalers, men who brought the entire world back to New Bedford on their whaleships through the curios they gathered.

The Field Museum turned that argument on its head, suggesting that the material objects were necessary because the whaling industry had done tremendous *harm*—driving the whale species to the brink of extinction. The museum hammered the point home in the next sentence: "The introduction of the modern harpoon fired from a gun having taken the place of the hand lance, is devastating the sea in a manner somewhat similar to the slaughter of the denizens of the prairie by the repeating rifle in the hands of the modern hunter."[18] The whaling industry had long endured monikers like "decaying," "dying," "withering," etc. When *Harper's Weekly* wrote of the *Progress*'s trip to

The Last Voyage of the Whaling Bark *Progress*

Chicago back in 1892, the article of three short paragraphs managed to cram in the words "old-fashioned," "by-gone days," "rotting," "deserted," and "obsolete."[19] "Decadence" was also familiar, but in this case, it seemed to go further than just the decay of the industry. It also served as a judgment that whalers had overhunted their prey in the pursuit of profits and gain. It was effectively an indictment of New Bedford.

Visitors to Hall 23 were treated to one of the *Progress*'s whaleboats in the center of the room "fitted out ready for service, and containing life-size figures of six sailors." Along the walls and throughout the room were cases and objects from the *Progress*, the now-familiar mashup of objects from the whaling industry; curios collected from around the world; pieces of marine species such as teeth, horns, and jaws; and objects that simply had a nautical theme, including the quadrant from the *Mayflower*.

However, there are hints of pedagogy relating to whaling that the *Progress* had never seemed able to achieve on its own. For example, Case 16 featured "A collection of implements used in whaling: guns, harpoons, and bomb-lances of various kinds, illustrating the progress of the whaling industry." When G.A. Coffin had drawn the same implements in 1892 they were seen fanned over each other in an attractive spray, a bouquet-like display, their variety of shapes highlighted more than their function or purpose. Now the Field Museum was breaking apart the bouquet and arranging them in a chronological, teleological way to emphasize the idea of technological progress. Perhaps not surprisingly, it was completely reminiscent of the Smithsonian whaling exhibits back in the late 1880s, rather than what the *Progress* had actually presented. The curatorial "search for order" once again reigned supreme. In another corner of the gallery an entire display was dedicated to the ship's cooper and his tools, a topic of vital importance to the whaling industry but never mentioned in the *Progress*'s materials. Whereas the State Street Bridge brochure had glossed over the process of rendering blubber in the tryworks, the Field Museum actually included the whaling terminology of "the blanket piece" in their guidebook.[20]

The attention to the *Progress*'s collections, including the thorny message about whaling as a worthy subject of display, lasted less than a year. By 1895 and the third edition of the *Guide to the Field Columbian Museum*, the "Section of Fishery Industries"—which had been entirely comprised of *Progress* objects—had been eliminated. Hall 23 was closed and was in the process of being turned over to the Zoology Department for its osteological collection, i.e., skeletons. While the case of barometers, log books, and quadrants from whaling had survived, the displays of other whaling implements including the progressive display of harpoons, were gone, as was the fully equipped

Ten. New Bedford, 1904

whaleboat and mannequin crew.[21] It was as if the stink of whaling had been detected and was being scrubbed out of existence.

The second annual report of the Museum hinted at the changes that had been afoot over the past year: "The system of the Museum generally has been much improved during the past twelve months.... While the work of the first year may be characterized as rushing and spasmodic, the labor of the second year has been even and steady." Rushing and spasmodic are hardly ringing endorsements for a year's worth of curatorial work. The report would add for good measure, "Much of the crudeness of the original installation has disappeared during this year, producing greater harmony of method, and connecting divisions with closer regard for an intelligent and comprehensive scheme of installation."[22] Hall 23, though not singled out, was clearly one of those original installations guilty of crudeness, and perhaps worse.

It turned out that the *Progress* collection had likely become one of several pawns in the museum's tumultuous first year, a fact that is barely concealed in the aforementioned Director's report. Director Frederick J.V. Skiff was a former Columbian Exposition administrator with no scientific or museum training. To many of his staff, he was a sycophant who favored the whims of the museum's wealthy patrons over scientific research and inquiry. By 1895 he had already endured several high-profile departures from his departments of Anthropology and Zoology. Skiff also served as the chief of the Department of Industrial Arts, under which the whaling gallery fell. Hall 23 was essentially his baby. As the battles with Skiff's major departments and curators unfolded, it was easy for him to offer the whaling display up as a sacrificial lamb to secure some good will. This meant cutting the "Section of Fishery Industries" from the museum's plans, clearing out many of the whaling artifacts, and offering the physical space to his internal critics.

Of course, not everything that came off the *Progress* was *persona non grata* to the Field Museum's curatorial staff. The Zoology Department happily took a long list of jaws, teeth, skulls, tusks, ribs, and even "7 penes of walrus." Many of the "Southern Seas" grass mats, belts, and skirts that so many visitors and reporters had commented on found a home in the Anthropology Department. Even the Botany and Geology Departments found a few random items to add to their collections.[23] As for the whaling harpoons, lances, etc., they too turned up around the museum from time to time—an alcove here, a wall there. But the overriding message was clear: If the whaling industry wanted memorialization and museum space dedicated to their venerable trade, they would need to look elsewhere. The entire Department of Industrial Arts was eliminated by 1896.[24]

Which is why a meeting in Pittsburgh in 1907 was so fortuitous. One

The Last Voyage of the Whaling Bark *Progress*

of the Field's curators of anthropology had, while in the Steel City, met with representatives from the Peabody Museum in Salem, Massachusetts (today the Peabody Essex Museum). These officials were quite interested in taking the whaling objects off the hands of the Field "for installation in their rapidly growing, and already famous, marine museum."[25] The Field wasted no time following up on the curator's new lead. A letter from the Director's Office put it as plainly as possible: "If you will kindly advise me how you wish this material shipped, it will be prepared and forwarded to you at once."[26] Director Skiff was true to his word. It took just three months for five boxes and four crates to be loaded upon the Michigan Central Railroad and shipped via fast freight from Chicago to Salem. The gift from the Field to the Peabody was a compendium of whaling industry artifacts—harpoons, lances, guns, spears, charts, logbooks, models, scrimshaw, compasses, quadrants, and even an anchor and a cannon. The railroad listed the total weight of the transferred objects at 2,600 pounds.[27] Over a ton of the whaling industry's tools were returning to Massachusetts, having received final marching orders from a city that never quite knew what to do with them.

Thus, it was the fate of the objects *in* the *Progress* to be spread among two major museums in both Chicago and Salem. The fate *of* the *Progress* herself was something else entirely. At the same time Weaver sold the *Progress*'s collection to the Field for $5,000 he appears to have also made a side deal with J.K. Peterson, "employed at the museum," to purchase the bark for $300. It was an Icarus-worthy tumble from the $3,000 Weaver likely paid for the whaling bark. The deal was also predicated on Peterson retaining the good will of Jackson Park officials to allow the *Progress* to remain in the South Pond while he decided what to do with her.[28]

The whaleship had miraculously managed to avoid injury from massive fires that had wiped out most of the fair's abandoned structures in January and July 1894. Survival from the second fire was particularly surprising given that the nearby Agricultural Building had been reduced to rubble by the conflagration. By late 1894 articles were focusing in earnest on the *Progress*'s abandonment, "deserted and forlorn in the South Pond." "The rats have abandoned it," the *Tribune* piece continued, "and it is visited now only by the occasional idler, curiosity-seeker, or tourist … there is no more progress for the Progress. Its day is dead."[29] There were murmurs in 1895 that Peterson might have a junk buyer interested in her copper hull, but the deal appears to have fallen through.[30]

By 1896 the situation had become something of a joke. "Does the South-Sider who made a bet of $5 that the old whaler Progress in Jackson Park will outlast the Statue of the Republic expect to live to see the wager decided,"

Ten. New Bedford, 1904

asked the *Tribune*.[31] Things came to a head in July of that year when the *Progress* was finally given her eviction notice. "Few frog ponds have been honored as has been the one near the east end of the old court of honor in Jackson Park," wrote the *Chicago Chronicle*. "The famous old whaling ship Progress has been in it ever since the opening of the World's Fair, and it is today stuck in the muddy bottom of the shallow little pool.... The ship is careened over to one side, its interior is desolated and barren and the ropes and cording in the rigging are rotting and almost ready to fall into pieces."

Still, the *Chronicle* reported, there was happy news: "A few days ago the Progress was sold to an enterprising buyer, who is to put her on her sea legs once more. An expert diver is now at work, calking [sic] the yawning seams in the sides of the boat, and it is probable that the mothers of Hyde Park will not have to worry about the presence of the Progress many weeks longer."[32] The *Tribune* added further details that the purchasers were the Newton Brothers—a family of successful milk distributors who had originally hailed from Massachusetts—and that they would refit her for a trip to New York where she would be a "training ship for merchant seamen."[33] One wonders at the connection between a decrepit whaler and milk moguls, although the fair's dairy pavilion had been immediately adjacent to the whaling museum. Perhaps the Newton Brothers had spent six months gazing out at the old bark. "Whaler Coming Back from Chicago," the *New York Times* announced in their July 13, 1896, edition while the *Chicago Record* promised the Newton Brothers would "put her in good shape again."[34]

There was, however, a serious logistical problem. In the nearly three years since the fair had closed, the entrance to the South Pond had silted up. "The entrance to the south pond was so filled with drifting sand," reported the *Inter Ocean*, "that the old bark had to be hauled over the tongue of land that divides the pond from the lake." Rollers had to be placed under the bark and it took two tugboats pulling from the lake side, in a "supreme test of tug power," to haul the whaler "bodily across the land into the lake."[35] The bark made it to the mouth of the Calumet River in South Chicago, where the refitting work was to be undertaken. Unfortunately, whatever patchwork had been done to get to her that point failed, and she sank again, this time in the Calumet's foul discharge into Lake Michigan. The *Tribune* offered a parting jab: "Frequenters of Jackson Park assert that the expression on the face of the Statue of the Republic has become noticeably sadder since the removal of the old whaler Progress from the south pond, and that early in the morning, something suspiciously resembling tears may be seen stealing down the large and dingy nose of the lonely figure that still watches over the ruins of the Court of Honor."[36]

The Last Voyage of the Whaling Bark *Progress*

Whatever plans there were for refitting the *Progress* and turning her into a training ship seemed to die at that point. The *Progress* had settled into the river mud and looked fated to remain there. By June 1897 Peterson was advertising the *Progress* for immediate sale in the classified ad pages of the *Chicago Tribune*. His Saturday advertisement fell on the page just below the classified "Wanted—A well trained driving goat."[37] Another round of speculation that the hulk might be sold stirred again October 1897, but by now the pattern was familiar, and the prospects for anything happening with the derelict ship faded quickly. By Christmas, 1901, shipping interests were complaining that the decaying hulk was in the way of river traffic, a claim reminiscent of her time at the State Street Bridge. In the final days of 1901, her snow-covered remains seemed destined to be blown apart by dynamite, clearing the way for the channel to be deepened and for large vessels to proceed upstream.[38] But this plan too appears to have lingered on without completion.

It would take all the way until a frigid night on February 25, 1902, for a mysterious fire (or "a red wind, a gale of fire that she could not weather" if you were writing for a Hearst newspaper) to engulf the wreck and set off a final round of reporting: "As the flames shot high from the oil-soked [sic] timbers a crowd gathered on the river bank to view the sight. No steps were taken to extinguish the fire, which soon consumed the vessel to the water's edge. The remnants of the old whaler soon will be blown to pieces by dynamite, as the fire left enough of the wreck to prove a menace to navigation."[39]

Just a few weeks before, the library and nascent historical society in Westerly, Rhode Island, received a note and a mysterious package. Enclosed was a heavy, wooden, intricately carved and scrolled billethead (so called when it is a non-representational version of a ship's *figure*head, perhaps most commonly associated with carved depictions of busty mermaids and maidens). The note read: "Dear Sir—I send you today the figurehead of the old whaling ship Charles Phelps, later renamed the Progress, which is just being broken up here." The author was E.C. Greenman, the grandson of one of the whaleship's original builders, and he hoped the people of Westerly would welcome the bit of history that the billethead represented. He also enclosed photographs of wreck, from which he had saved the sculptural element.[40] Later that winter the entirety of the *Progress* would be lost to fire, dynamite, and river water. The rescued billethead remains on display at the Westerly Library today.

It all made for an interesting story. Yet it was one of thousands of such postscripts to the fair about the fate of former pavilions, favorite exhibits, and colorful characters from the Midway. Given that the fair covered over 600 acres of land with countless ways to awe and entertain, these kinds of stories

Ten. New Bedford, 1904

Couple aboard the *Progress* in the Calumet River. Photograph by E.C. Greenman, circa 1902. The couple in this photograph are likely E.C. Greenman and his wife. It was taken as he was rescuing the *Progress*'s billethead prior to her final demolition (courtesy of Westerly Library & Wilcox Park).

sprinkled the newspapers and periodicals of America for years after the Columbian Exposition.

Except, of course, the fate of the *Progress* was much more than an interesting postscript because it was a narrative that so easily metastasized into a larger metaphor for the continued decline and decay of whaling. That fate was sealed the moment the New Bedford Board of Trade sold the ship in order to become a representation of whaling at the Columbian Exposition. "It was thought that the vessel's great past and the curiosities and interesting relics which were stored on board would prove a splendid drawing card and a most profitable venture," the *Chicago Chronicle* reminded readers.[41]

But even more than Henry Weaver and the other stakeholders in the Arctic Whaling Exhibit Company, it was New Bedford which had gambled and lost. In 1897 the city hosted semi-centennial celebrations to which President McKinley telegraphed congratulations. One of the city's leaders, George E. Briggs (a cotton man, not a whaling man), telegraphed back to the president a reminder of the city's *Lucem Diffundo* motto, adding that New Bedford "reviews her past with pride and anticipates the future with hope."[42] Except the *Progress* had been a referendum on the power of whaling to excite and inspire the American public, and the answer had been a resounding "no."

Torn from their community of origin and sent a thousand miles away, neither the ship's history nor the curiosities and relics aboard were of interest to fair audiences. From the Great Lakes to the Chicago River to the Calumet Harbor the *Progress* had been an obstruction to the flow of modern life. At the Columbian Field Museum, the objects were given a brief screening to the public and then were hurriedly ushered out the door. "The burning of the ancient whaler Progress, nearly a century old," remarked one author, "which has been moored since the World's Fair in the waters of South Chicago harbor, shows how little sentiment has to do with a bustling community wedded to business."[43] And the citizens and leaders of New Bedford knew and followed all of this, because it kept reappearing on their doorstep.

Just as Chicagoans were checking in on the *Progress*'s increasingly decrepit state year after year, so too were the New Bedford newspapers following the story. They picked up word in 1895 that perhaps another fate for the old whaler could be use by the Illinois Naval Reserve as a training ship. The scheme failed to materialize. They followed the exodus of the *Progress* out of the South Pond in the summer of 1896. They followed her sinking and continued decay at the mouth of the Calumet River.[44] But perhaps most importantly they followed how the whole story was being seen by the outside world. Nothing did a better job of that than a piece of writing out of Chicago in 1899 that not only landed in New Bedford, but Westerly, Rhode Island, where the *Charles Phelps* nee *Progress* was originally built and launched.

Originating with the *Chicago Daily News*, the article recounted both the ship's proud history and her trials and tribulations in Chicago. It then switched to a bit of storytelling, opening with: "The other day an old gray-haired man walked across the bridge near by and strolled over to where she lay. He looked at her in puzzled surprise. He climbed to her deck and inspected the cabin. 'Why, ef'taint the ole Progress,' he said, half to himself." The man in the story is then transported back to his own youth, while the wreck around him disappears into an imagined bark, "a handsome craft," in the bloom of her prime whaling days. He continues seeing phantasms of spotting whales and hearing "there she blows" before eventually being pulled out of his reverie. "The old man sat down on a broken timber and buried his face in his hands. The wind stirred through the rigging of the Progress, and a broken shroud hanging from the mizzenmast brushed against his cheek. He started up and caught the

Ten. New Bedford, 1904

rope in his hand. 'Poor old girl,' he said. 'She feels as bad as I do. All alone, and a thousand miles from salt water, and both of us stranded.'"[45]

It is perhaps debatable whether those born and raised in Chicago—"a thousand miles from salt water"—necessarily saw the entirety of the metaphorical implications contained in this little fictional detour. But surely the words were not lost on the whaling communities who read them in reprintings, especially residents like William W. Crapo and Ellis Howland. It was the tale's grey-haired whaleman and the decaying whaleship that were "stranded and alone." That indictment reflected directly on the dying whaling industry anchored and rotting right before the eyes of Crapo and Howland, and right in their own backyard. "She feels as bad as I do," the whaler cried. Whaling had proved unwanted and unloved once already. Would New Bedford continue to turn away from the discomfort created by this failure? And so it came to be that less than ten months after New Bedford read about the final conflagration of the *Progress*, Ellis Howland stood at a podium and demanded that New Bedford create a museum to whaling. It was time for New Bedford to retake control of their narrative about the importance of whaling's history, more than ten years after they had sold it off to a syndicate from Chicago.

William Crapo would prove instrumental in securing a permanent museum to whaling beginning in 1906. Expanding from a donated bank building on Water Street, the enterprise would in time evolve into the world-famous New Bedford Whaling Museum—today a vibrant center of whaling history, science, and education. Crapo also personally commissioned and paid for a large, bronze whaling statue situated in front of the public library, "where hundreds of people passing every day could be reminded of the rugged sailors who made New Bedford possible."[46] The price tag of $25,000 was half of the Arctic Whaling Exhibit Company's capitalization cost. Unveiled in 1913, one of the library's trustees noted that everything about Crapo's statue was "correct historically," and that "the boat was produced from one that has had actual service, and may be found in the rooms of the Old Dartmouth Historical Society. It was measured and photographed and drawings of it were made under the artist's directions in the interest of exactness. The same museum furnished also the original of the harpoon."[47] The lavish attention to detail and historical correctness were a far cry from claims of a Fiji King and Windy City schooner sailors dressed up as whalers. It seemed, at last, New Bedford was getting an accurate depiction of whaling.

By the time the statue's coverings fell away in 1913 and revealed the shirtless harpooner to the gathered audience, many of those associated with the *Progress* were gone. Her New Bedford owner George F. Bartlett, who had sold the *Progress* to Henry Weaver, had died in 1905. His biography in a local 1899

The Last Voyage of the Whaling Bark *Progress*

Whalemen's statue unveiling exercises outside of the New Bedford Free Public Library, 1913. Mayor Charles S. Ashley is seen speaking. Also seated at the podium is William Crapo. Throughout the dedication ceremony speakers noted the historical accuracy and attention to detail that had been put into the statue, including the fact that it was modeled from actual whaling implements (courtesy of the New Bedford Whaling Museum).

publication noted, "It will be remembered that this firm owned the famous old ship Progress, which figured in the 'Arctic disaster of 1871,' sailed with the 'Stone Fleet,' and was finally sold by Mr. Bartlett to Henry E. Weaver of Chicago for the World's Columbian Exposition, where it was exhibited."[48] Ironically, even in 1905 the debate continued as to whether New Bedford whaling was truly dead. "New Bedford Wharves Once Again Alive with Whalermen" announced one headline from that year, adding, "Whether the whaling industry is or is not in process of revival no mere outsider can say.... What is apparent, however, to the landlubber is that something is doing in the whaling fisheries just now."[49]

Henry Weaver had also died in 1905, just a few months after Bartlett, succumbing to a sudden stroke. Prior to that, he had remained active in Chicago, eventually expanding his business interests into ice and

Ten. New Bedford, 1904

electricity, and his philanthropic interests to free public bathing facilities on Lake Michigan, and medals for heroic lifeguards who patrolled Chicago's waters.[50] The court cases surrounding the debts of the Spectatorium would drag on well after his death. Steele MacKaye, meanwhile, had died on a train bound for California back in 1894. Given the ultimately strange intersection between a midwestern coal baron and a New Bedford whaleship, there was a certain cosmic synergy to a fact reported by a Pennsylvania newspaper in 1902: "After diligent search you find the one or two houses that do make a business of whale products, and you learn that there is now just one staple use for sperm oil—miners' lamps."[51]

Fitch Crane would stay on in Chicago through 1896, a period that included charges brought by the Humane Society of abusing a horse. Crane would return to upstate New York by 1897 where he presented the *Progress*'s sextant to a friend and editor of the *Wellsville Democrat*, Adolphus Cowles. He would later move on from horses to focus on automobiles and by 1913 would be dubbed Horseheads, New York's "automobile king." He died in 1927.[52]

Captain "Bloody Dan" Gifford had continued to serve as a whaling captain for the rest of his life. However, his susceptibility to illness continued to plague him. He last had the helm of a whaler called the *Sunbeam*, which had departed New Bedford on September 7, 1897. Two years later in September 1899 he was in the Azores. There he had to give up his command to a Captain Reynolds and return to New Bedford due to illness. He died back home in New Bedford on December 10, 1899. His voyage on the *Gayhead* to Antarctica had not revitalized the whaling industry; however, he did report back that he had taken 15 whales and "filled every cask" on the bark. His report came just a few weeks before the Columbian Exposition opened for business and the *Progress*—the museum to the dying industry of whaling—went on display.[53]

Only Captain James Dowden was alive to witness William Crapo unveil his monumental statue to whaling; however, if he was in the audience, he went unacknowledged by the speakers who assembled that day. Perhaps this was because Dowden had recently come out against continuing the whaling trade, suggesting that "the United States and Canada join in an agreement to prevent further indiscriminate killing of whales."[54] He died in 1915. It was noted in Dowden's obituary that the young 12-year-old that he had brought aboard the *Progress* during the Arctic rescue—William Fish Williams—had grown up to become chief engineer of New Bedford's Harbor and Land Commission.[55]

Then, in 1939, a group of New York City investors sent a letter of inquiry to the mayor of New Bedford. A ship brokerage firm representing the group

The Last Voyage of the Whaling Bark *Progress*

wanted to purchase the old whaler *Charles W. Morgan*, by now the last of the wooden ships from America's whaling fleet. The idea was to take the ailing ship, which had barely survived a New Bedford hurricane strike the year before, and put her "into first-class condition" for the intent "to exhibit the boat the World's fair in New York. The offer specified that the whaler would be returned in six months."[56] Nearly fifty years later, history was repeating itself—a group of investors and outsiders with no connection to whaling wanted to take a whaleship from New Bedford, fix her up, and put her on display at a World's Fair that would ultimately attract nearly a third of the nation. Except this time, the offer met a chilly reception. "I am opposed to the plan," said William H. Tripp, a curator for the Old Dartmouth Historical Society and trustee of the *Morgan*'s present home at Round Hill, South Dartmouth, "because I have in mind what happened to the bark Progress when she was towed to the Chicago World's Fair in 1893. She never came back. She started to rot in her berth and then burned to the waterline."[57]

Instead the *Charles W. Morgan* left New Bedford in November 1941 and headed to her new home in Mystic, Connecticut, to what would later be known as Mystic Seaport. The recalcitrant William H. Tripp, who had opposed the fair offer, would serve as her master, and the crew included Everett Allen, a young journalist from New Bedford's *Standard-Times*. Allen would write during the voyage, "She has come alive again, with a white ripple under her forefoot and bubbling water swirling around her quarters. Her life has returned, from truck to keelson, for this brief resurrection. Mariner or landlubber, we all feel it."[58]

What would have happened if the *Morgan* had gone to the New York World's Fair? What if Tripp, like so many in New Bedford, had ignored or forgotten the sad story of the *Progress*? Would we still have the one, remaining, precious whaling ship that is on display at Mystic Seaport today? Everett Allen ended his piece describing the sailing of the *Morgan* to Mystic with these words: "it has been an experience that never will be repeated … and I have a hunch our grandfathers—and great grandfathers—are glad we had this opportunity. Somehow, it's been like meeting them."[59] In the end, it was only by avoiding another world's fair—and by heeding the histories and lessons from the *Progress*'s strange and twisted journey—that the ghosts of America's whaling industry had, at long last, discovered their true, authentic voice.

Chapter Notes

Preface

1. "Digitized Logbooks," New Bedford Whaling Museum Research Library, accessed October 22, 2019, https://www.whalingmuseum.org/explore/library/logbooks-journals.

2. Barbara Kirschenblatt-Gimblett, *Destination Culture: Tourism, Museums, and Heritage* (Berkley: University of California Press, 1998), 3.

Chapter One

1. Account reconstructed from: G. A. Coffin, *"There She Blows" or the Story of the Progress* (Chicago: The Arctic Whaling Expedition Company, 1893); "The Progress," *New Bedford Mercury*, June 8, 1892; "The Progress Sails," *Evening Standard* [New Bedford], June 8, 1892; Joe Silvia, "Historical Personages of New Bedford: James E. Reed," *Newbedfordguide.com*, March 9, 2013, https://www.newbedfordguide.com/james-e-reed/2013/03/09.

2. Account reconstructed from: "End of Two Historic Vessels," *Chicago Daily Tribune*, July 18, 1897; "Madagascar in Peril," *Chicago Daily Tribune*, November 3, 1899; "Dynamite for an Old Whaler," *Chicago Daily Tribune*, December 21, 1901; Christopher Thale, "Calumet River System," *Encyclopedia of Chicago*, Chicago Historical Society, accessed October 22, 2019, http://www.encyclopedia.chicagohistory.org/pages/203.html. Weather and lunar information reconstructed from National Weather Service historical weather data, accessed on October 22, 2019, http://w2.weather.gov/climate/xmacis.php?wfo=lot; Lunar information from moonpage.com.

3. Ryan S. Burg, "Rebuilding the Iron Cage: Post-Failure Organizing in Newspapers and Investment Banks" (Ph.D. diss., University of Pennsylvania, 2011), 75.

4. Ronald R. Brown, *Dying on the Job: Murder and Mayhem in the American Workplace* (New York: Rowman and Littlefield Publishers, 2013), 229.

5. William Henry Seward, *The Whale Fishery, and American Commerce in the Pacific Ocean. Speech of William H. Seward, in the Senate of the United States, July 29, 1852* (Washington, D.C.: Buell & Blanchard, 1852), 3.

6. Arnold Lewis, *An Early Encounter with Tomorrow* (Urbana: University of Illinois Press, 2001), 178.

Chapter Two

1. "Municipal Affairs," *New Bedford Mercury*, June 25, 1847.

2. "Town Seals Symbolize History: New Bedford," unattributed newspaper clipping, Old Dartmouth Historical Society Scrapbook, #1, 145, New Bedford Whaling Museum Research Library.

3. For additional background information on the city seal, see Arthur P. Motta, Jr., "Seal of the City of New Bedford, Massachusetts," City of New Bedford Office of Tourism & Marketing, 2013. "Oil City" referenced in William Kirk, ed., *A Modern City: Providence, Rhode Island and Its Activities* (Chicago: University of Chicago Press, 1909), 65.

4. Information from James T. Almy census records: "James Almy," 1850 United States Federal Census, City of New Bedford, Massachusetts, 269; "James Almy," 1860 United States Federal Census, City of

Notes—Chapter Two

New Bedford, Massachusetts, 191; "James T. Almy," 1865 Massachusetts State Census, City of New Bedford.

5. "New Bedford, Massachusetts," *The National Magazine and Industrial Record* 1, no. IV (1845): 328–343.

6. "Revised Ordinances of the City of New Bedford, 1853…Seal of the City," in *City Documents, New Bedford 1854–1855* (New Bedford: Benjamin Lindsey, City Printer, 1855), 3.

7. For example, see Marc Songini, *The Lost Fleet: A Yankee Whaler's Struggle Against the Confederate Navy and Arctic Disaster* (New York: St. Martin's Press, 2007).

8. Peter Feinman, "Whaling and Abolition: A Sample 'Path Through History,'" Institute of History, Archeology and Education, April 29, 2013, https://ihare.org/2013/04/29/whaling-and-abolition-a-sample-path-through-history/; "Librarian's Report," in *The Fiftieth Annual Report of the Trustees of the Free Public Library of the City of New Bedford* (New Bedford: A.E. Coffin Press, 1902), 9.

9. Zephaniah W. Pease, *The Centenary of the Merchants National Bank* (New Bedford: Reynolds Printing, 1925), 7.

10. Mike Vogel, "The Light from the Whale," *The Keeper's Log*, Fall, 2014, 56–62; "Beyond the Arctic Circle," *The Contributor*, January, 1895, 139. David Jaffee, *A New Nation of Goods: The Material Culture of Early America* (Philadelphia: University of Pennsylvania Press, 2010), 319.

11. "Rich and Oily," *The Daily Journal* [Wilmington, NC], July 12, 1855. Reprinted from *Boston Chronicle*.

12. Herman Melville, *Moby-Dick* (Pleasantville, NY: The Reader's Digest Association, 1989), 302. This Reader's Digest edition contains the complete text of Herman Melville's *Moby-Dick: Or, the Whale*, first published in 1851.

13. Philip Hoare, *The Whale: In Search of the Giants of the Sea* (New York: ECCO, 2010), 96.

14. As quoted in Eric Jay Dolin, *Leviathan: The History of Whaling in America* (New York: W. W. Norton, 2007), 113.

15. Olmsted as quoted in Stephanie McCurry, *Masters of Small Worlds: Yeoman Households, Gender Relations, & the Political Culture of the Antebellum South Carolina Low Country* (New York: Oxford University Press, 1995), 73.

16. Campbell Morfit, *A Treatise on Chemistry Applied to the Manufacture of Soap and Candles* (Philadelphia: Parry and McMillian, 1856), 504. See also Vogel, "The Light from the Whale," 56–62.

17. "Oil," *Whalemen's Shipping List and Merchants' Transcript*, May 2, 1843. Reprinted from the *Nantucket Inquisitor*.

18. Treasury official quote: *Report of the Officers Constituting the Light-House Board* (Washington, DC: A. Boyd Hamilton, 1852), 272; "Sealed for Proposals," *New York Times*, January 10, 1854. This article contained the RFP for the annual oil purchase from the Treasury Department.

19. Dolin, *Leviathan*, 105; Michael Paterson, *Inside Dickens' London* (Cincinnati: David and Charles Limited, 2011), 17; "Streetlamps," *Old House Journal*, July/August, 1988, 52; Catherine M. Hintze and Robert Mulero, *The History and Design of New York City Streetlights, Past and Present* (Pittsburgh: Dorrance Publishing Co., 2017), 6.

20. Margaret S. Creighton, *Rites & Passages: The Experience of American Whaling, 1830–1870* (Cambridge: Cambridge University Press, 1995), 25.

21. Scott Institute for Energy Innovation, "What Fuel did Pittsburgh's First Street Lamps Consume?" Interview with Joel Tarr, *Energy Bite*, podcast audio, April 4, 2017, https://energybite.org/page/2/; Peggy Scott Laborde and John Magill, *Canal Street: New Orleans' Great Wide Way* (Gretna, LA: Pelican Publishing Company, 2006), 91; "Historic Warm White Street Lights in Washington," *Capitol Hill Restoration Society*, March, 2017, http://chrs.org/wp-content/uploads/2017/03/StreetLghtsWashingtonHistory4-1-2017eap.pdf.

22. *City Documents, The Mayor's Address* (New Bedford: Benjamin Lindsey, 1852), 7.

23. Melville, *Moby-Dick*, 317.

24. Michael P. Graves, "Functions of Key Metaphors in Early Quaker Sermons, 1671–1700," *Quarterly Journal of Speech* 69, no. 4 (1983): 366.

25. Arthur R. Roberts, "A Quaker Understanding of Jesus Christ," *Quaker Religious Thought* 93, no. 4 (1999): 5.

26. Robert A. Greene, "Whichcote, the

Notes—Chapter Three

Candle of the Lord, and Synderesis," *Journal of the History of Ideas* 52, no. 4. (1991): 639.

27. Sarah Lelia Crabtree, "A Holy Nation: The Quaker Itinerant Ministry in an Age of Revolution, 1750–1820" (Ph.D. diss., University of Minnesota, 2007), 84; 105.

28. "Words of Eloquence," *The Evening Standard*, October 11, 1897.

29. Earl F. Mulderink III, "'A Burning and Shining Light': Prosperity and Enlightened Governance in Antebellum New Bedford," in *New Bedford's Civil War* (New York: Fordham University, 2012), 10–31.

30. Hoare, *The Whale*, 67.

31. As quoted in Leonard Bolles Ellis, *History of New Bedford and Its Vicinity* (Syracuse: D. Mason & Co., 1892), 303.

32. Ephraim Peabody, *A Discourse Delivered at the First Public Meeting of the New Bedford Orphan's Home* (Boston: Isaac R. Butts, 1842), 20.

Chapter Three

1. "Lady Whalers," *Whalemen's Shipping List and Merchants' Transcript*, February 1, 1853.

2. The events in this chapter come primarily from Eliza Azelia Williams, "The Voyage of the *Florida*, 1858–1861: From the Journal of Eliza Azelia Williams," in *One Whaling Family*, ed. Harold Williams (Boston: Houghton Mifflin Company, 1964). Additional notes: "Interpretive Themes: Unique Cultures," New Bedford Whaling National Historical Park, accessed April 15, 2019, https://www.nps.gov/nebe/learn/historyculture/cultures.htm; Melville, *Moby Dick*.

3. Williams, "The Voyage of the *Florida*," 5–6.

4. Melville, *Moby-Dick*, 152.

5. Captain H. A. Chippendale, *Sails and Whales* (Boston: Houghton Mifflin, 1951), 94.

6. Williams, "The Voyage of the *Florida*," 21.

7. Nelson Cole Haley, *Whale Hunt* (New York: Ives Washburn, Inc., 1948), 74.

8. James McGuane, *The Hunted Whale* (New York: W. W. Norton & Company, 2013), 16.

9. Henry T. Cheever, *The Whale and his Captors* (New York: Harper and Brothers, 1864), 138; Melville, *Moby Dick*.

10. James Temple Brown, "Stray Leaves from a Whaleman's Log," *Century Illustrated Magazine*, February 1893, 516–517.

11. "'There she blows' and how 'she gets it,'" *Chicago Daily Tribune*, May 17, 1903.

12. Robert Ferguson, *Harpooner* (Philadelphia: University of Pennsylvania Press, 1936), 164–165.

13. Nancy Shoemaker, *Living with Whales: Documents and Oral Histories of Native New England Whaling History* (Amhurst: University of Massachusetts Press, 2014), 104.

14. William Henry Giles Kingston, *Peter the Whaler* (New York: E.P. Dutton & Co., 1906), 203.

15. Williams, "The Voyage of the *Florida*," 20–21.

16. Ibid., 10.

17. Ferguson, *Harpooner*, 24.

18. Williams, "The Voyage of the *Florida*," 22.

19. Melville, *Moby-Dick*, 271.

20. Williams, "The Voyage of the *Florida*," 26.

21. Granville Allen Mawer, *Ahab's Trade: The Saga of South Seas Whaling* (New York: St. Martin's Press, 1999), 69.

22. Williams, "The Voyage of the *Florida*," 25–26.

23. Ibid., 26.

24. Sturgeon Stewart, "The Whale and his Haunts, Article II," *The Canadian Magazine*, October 1909, 516.

25. Dolin, *Leviathan*, 86–88.

26. Frances Allyn Olmsted, *Incidents of a Whaling Voyage* (New York: D. Appleton and Co., 1841), 68; see also James Williford, "Whaling the Old Way," HUMANITIES: *The Magazine of the National Endowment for the Humanities* 31, no. 2 (2010): https://www.neh.gov/humanities/2010/marchapril/feature/whaling-the-old-way.

27. Haley, *Whale Hunt*, 110.

28. Creighton, *Rites & Passages*, 53.

29. Roger Starbuck, *The Golden Harpoon or, Lost Among the Floes: A Story of the Whaling Grounds* (New York: Beadle and Adams, Publishers, 1865). Word search through http://dimenovels.lib.niu.edu/islandora/object/dimenovels%3A119#page/1/mode/1up.

30. Kingston, *Peter the Whaler*, 332.

Notes—Chapters Four and Five

Chapter Four

1. "David B. Kempton," in *Our County and its People: A Descriptive and Biographical History of Bristol County, Massachusetts* (Boston: The Boston History Company, 1899), 304; "Biographical Note," *Finding Aid for Inventory of the David Batchelder Kempton Papers*, New Bedford Whaling Museum Research Library, accessed October 22, 2019, https://www.whalingmuseum.org/explore/library/finding-aids/mss97.

2. Melville, *Moby-Dick*, 82.

3. Lance E. Davis, Robert E. Gallman, and Karin Gleiter, *In Pursuit of Leviathan: Technology, Institutions, Productivity, and Profits in American Whaling, 1816–1906* (Chicago: University of Chicago Press, 1997), 381–422.

4. Ibid., 411; 394.

5. Kempton's purchase history reconstructed through: Alexander Starbuck, *History of the American Whale Fishery* (Secaucus, NJ: Castle Books, 1989); Judith Navas Lund, *Whaling Masters and Whaling Voyages Sailing form American Ports, A Compilation of Sources* (New Bedford: Ten Pound Island Book Co., 2001).

6. Matthew Howland to James Stewart, 18 January, 1861, *Matthew Howland letter book, 1858–1879 (inclusive.)* Vol. 1. Mss: 252 1858–1879 H864. Baker Library, Harvard Business School, https://iiif.lib.harvard.edu/manifests/view/drs:48241696$1i.

7. For more on the Stone Fleet see: Mary K. Bercaw Edwards, "An Old Sailor's Lament: Herman Melville, the Stone Fleet, and the Judgment of History," *Leviathan* 9, no. 3 (2007): 51–64. Contemporary dollar value calculation made through https://www.measuringworth.com/calculators/uscompare/relativevalue.php.

8. Dolin, *Leviathan*, 311.

9. Matthew Howland to Edward Phinney, 5 January, 1862, *Matthew Howland letter book, 1858–1879 (inclusive.)* Vol. 1.

10. Muderink, *New Bedford's Civil War*, 150–151.

11. Account reconstructed from: "Samuel and Thomas, 1863" in Starbuck, *History of the American Whale Fishery*, 594–595. "The Pirate Florida," *New York Times*, July 13, 1864; "Destruction of the Bark Golconda of this Port, by the Rebel Steamer Florida," *Whalemen's Shipping List and Merchants' Transcript*, July 19, 1864; W. Craig Gaines, *Encyclopedia of Civil War Shipwrecks* (Baton Rouge: Louisiana State University Press, 2008), 21.

12. Statistics on *Shenandoah*'s run: Starbuck, *History of the American Whale Fishery*, 103; details of the *Waverly*: Lynn Schooler, *The Last Shot* (New York: Harpers Collins, 2005), 228.

13. Quote reprinted in "1865," *Evening Standard Special Anniversary Edition*, February 15, 1900.

14. Data on turnaround from port calculated from Starbuck, *History of the American Whale Fishery*, for ships that returned in port in 1864 or 1865 and sailed in 1865.

15. Matthew Howland to Paul Greene, 15 December, 1858; Matthew Howland to Valentine Lewis, 3 August, 1859; Matthew Howland to Edward Phinney, 5 January 1862, *Matthew Howland letter book, 1858–1879 (inclusive.)* Vol. 1.

16. "Ship Building," *Whalemen's Shipping List and Merchants' Transcript*, May 21, 1850.

17. Davis, et. al., 386–387.

18. Matthew Howland to Edward Phinney, 5 January, 1862, *Matthew Howland letter book, 1858–1879 (inclusive.)* Vol. 1.

19. Mary Kempton Taber, "The Kempton Family in Old Dartmouth," in *Old Dartmouth Historical Sketches*, no. 21 (1908), accessed on October 22, 2019, https://www.whalingmuseum.org/explore/library/publications/old-dartmouth-historical-sketches/odhs-number-21.

Chapter Five

1. Many sources have retold the story of the 1871 disaster. The most useful and frequently-consulted in writing this chapter were: Songini, *The Lost Fleet*; Dolin, *Leviathan*; William Fish Williams, "The Voyage of the *Florence*, 1873–1874," in *One Whaling Family*, ed. Harold Williams (Boston: Houghton Mifflin Company, 1964); Everett S. Allen, *Children of the Light* (Boston: Little, Brown, and Company, 1973); Martin W. Sandler, *Trapped In Ice!* (New York: Scholastic Non-Fiction, 2006). Although this last source is for a juvenile non-fiction market, it provides an account of events notable for its clarity and brevity.

Notes—Chapter Six

2. Matthew Howland to Valentine Lewis, 2 March, 1871; Matthew Howland to Robert Jones, 3 March, 1871, *Matthew Howland letter book, 1858–1879 (inclusive.)* Vol. 1.

3. "Introduction of Gas into New Bedford," *Whalemen's Shipping List and Merchants' Transcript*, February 8, 1853.

4. As quoted in Donald Warrin, *So Ends This Day: The Portuguese in American Whaling 1765–1927* (East Providence, RI: Signature Printing, 2010), 169–170.

5. Ellis, *History of New Bedford and Its Vicinity*, 421.

6. Starbuck, *History of the American Whale Fishery*, 555.

7. Williams, "The Voyage of the *Florence*," 235–236.

8. "Occasional Notes," *The Pall Mall Budget*, December 6, 1879, 17.

9. As quoted in John R. Bockstoce, *Furs and Frontiers in the Far North: The Contest among Native and Foreign Nations for the Bering Strait Fur Trade* (New Haven: Yale University Press, 2009), 316.

10. "August 15, 1871" and "August 27, 1871," in *Logbook of the Lagoda (Bark) of New Bedford, mastered by Stephen Swift, on voyage from 25 July 1868 to 1 June 1873*, New Bedford Whaling Museum, accessed October 22, 2019, https://archive.org/details/logbookoflagodab00lago.

11. Williams, "The Voyage of the *Florence*," 235.

12. Mary L. Sherman, "The World's Fair," *Los Angeles Times*, September 11, 1893; *The Arctic Whaler Progress A Complete Marine and Whaling Museum* (Chicago: Knight, Leonard & Co., 1892).

13. "To Chicago Whaling," *Daily Mercury* [New Bedford], June 7, 1892.

14. "In the Ice," *San Francisco Chronicle*, November 7, 1871.

15. "Hero of Old-Time Whaling Rescue," *Boston Daily Globe*, March 10, 1911; "Hero of Arctic Ice Pack of 1871 Dead," *Boston Daily Globe*, April 6, 1915.

16. "In the Ice," *San Francisco Chronicle*.

17. "September 1, 1871," *Logbook of the Lagoda*.

18. *Owners and Crew of the Hawaiian Bark Arctic*, SR 577, 52rd Cong., 1st sess., *Reports of Committees of the United States Senate for the First Session of the Fifty-Second Congress*, Vol 4, 2.

19. "September 9, 1871,"; "September 10, 1871," *Logbook of the Lagoda*.

20. "September 11, 1871," *Logbook of the Lagoda*.

21. *Owners and Crew of the Hawaiian Bark Arctic*, 1; 12.

22. As quoted in Peter Nichols, *Oil and Ice: A Story of Arctic Disaster and the Rise and Fall of America's Last Whaling Dynasty* (New York: Penguin, 2009), 237.

23. *Owners and Crew of the Hawaiian Bark Arctic*, 8.

24. "September 13, 1871," *Logbook of the Lagoda*.

25. "Loss of the Arctic Fleet," *New York Times*, November 7, 1871.

26. "Hero of Old-Time Whaling Rescue," *Boston Daily Globe*; "Death of Capt. Dowden of New Bedford Recalls Arctic Whaling Disaster of 1871," *Boston Daily Globe*, April 11, 1915; "Hemmed in by Arctic Ice: The Story of Great Disaster Twenty Years Ago," *New York Times*, July 19, 1891.

27. Michael P. Dyer, "The ship Lagoda: The Maritime History of an American Icon, 1826–1890," *New Bedford Whaling Museum Blog*, March 28, 2016, https://whalingmuseumblog.org/tag/arctic-disaster-of-1871/; Elton W. Hall, "Panoramic Views of Whaling by Benjamin Russell," in *Old Dartmouth Historical Sketches*, no. 80 (1981), accessed on October 22, 2019, https://www.whalingmuseum.org/explore/library/publications/old-dartmouth-historical-sketches/odhs_no_80.

28. "Russell's Arctic Pictures," *Whalemen's Shipping List and Merchants' Transcript*, April 9, 1872.

Chapter Six

1. Information for this chapter about the Brennans and other Irish immigrants based on: "George Brennan," 1880 United States Census, New Bedford, Massachusetts, 56.

2. Kingston Wm. Heath, *The Patina of Place: The Cultural Weathering of a New England Industrial Landscape* (Knoxville: University of Tennessee Press, 2001), 52; 79.

3. Some surnames corrected from original census-taker's spelling to match conventional spellings.

4. "From These Strains—The Irish in

Notes—Chapter Six

New Bedford: History of Irish Catholic Churches, Excerpted from *The Evening Standard*, August 30, 1887, prepared by Etta F. Martin," *Newbedfordhistory.com*, November 2, 2017, http://www.newbedfordhistory.com/2017/11/02/from-these-strains-the-irish-in-new-bedford/.

5. Detail about the clock tower from: "Mill Notes," *American Wool and Cotton Reporter*, September 23, 1920, 66.

6. See Thomas Dublin, ed., *Farm to Factory: Women's Letters, 1830–1860* (New York: Columbia University Press, 1981); Joshua L. Rosenbloom, "The Challenges of Economic Maturity: New England, 1880–1940," *NBER Papers on The Program on the Development of the American Economy*, The National Bureau of Economic Research, February 1999, https://www.nber.org/papers/h0113.pdf; Zephaniah Walter Pease, *New Bedford, Massachusetts: Its History, Industries, Institutions, and Attractions* (New Bedford: Mercury Publishing Company, 1889).

7. "General sentiments" quote: Pease. *New Bedford, Massachusetts*, 456; Semi-Centennial quote: Robert Grieve, ed., *New Bedford's Semi-Centennial Souvenir* (Providence: Journal of Commerce Company, 1897), 18; "stumbling block" quote: Zephaniah Walter Pease, *History of New Bedford* (New York: The Lewis Historical Publishing Company, 1918), 212; Final quote: *Republican Standard*, April 8, 1869. Reprinted from *Fall River News*.

8. As quoted in: Rose Pearl Rodrigues, "Occupational Mobility of Portuguese Males in New Bedford, Massachusetts: 1870 to 1900" (Ph.D. diss., New School for Social Research, 1990), 98.

9. As quoted in: Thomas Austin McCullin, "Industrialization and Social Change in a Nineteenth Century Port City: New Bedford, Massachusetts, 1865–1900," (Ph.D. diss., University of Wisconsin, Madison, 1976), 16. There was also one failed mill during this period—the New Bedford Steam Company. Samuel Rodman wrote in his diary in 1846 ,"went on a disagreeable mission to see if I could by an exhibition of my statements induce some of the prominent men of our town to take an interest in the cotton mill … the proposal for meeting heavy liabilities by gaining new colleagues to the enterprise seems very faint." As quoted in Seymour Louis Wolfein, *The Decline of a Cotton Textile City* (New York: Columbia University Press, 1944), 149.

10. "The Arctic Disaster," *Whalemen's Shipping List and Merchants' Transcript*, November 14, 1871.

11. "A New Cotton Mill," *Whalemen's Shipping List and Merchants' Transcript*, June 6, 1871; "Potomska Mills" *Republican Standard*, October 5, 1871.

12. "Potomska Mills," *Republican Standard*, July 18, 1872.

13. *City Documents, The Mayor's Address* (New Bedford: Fessendon and Baker, City Printers, 1871), 4.

14. Untitled editorial, *Republican Standard*, March 3, 1870.

15. Heath, *The Patina of Place*, 39–44.

16. Barbara Clayton and Kathleen Whitley, *Guide to New Bedford* (Chester, CT: The Globe Pequot Press, 1979), 26.

17. Dolin, *Leviathan*, 354.

18. Heath, *The Patina of Place*, 70.

19. Charles B. Spahr, *America's Working People* (London: Longmans, Green, and Co., 1900), 6.

20. Walter Sheldon Tower, *A History of the American Whale Fishery* (Philadelphia: University of Pennsylvania, 1907), 74–75.

21. Elmo P. Hohman, *The American Whaleman*, reprint of 1928 edition (Clifton: Augustus M Kelley, 1972), 300.

22. "The Whale-Fishery," *Whalemen's Shipping List and Merchants' Transcript*, November 21, 1871.

23. Nichols, *Oil and Ice*, 261; Joseph Martin Butler, "J. & W. R. Wing of New Bedford: A Study of the Impact on a Declining Industry upon an American Whaling Agency" (Ph.D. diss., Pennsylvania State University, 1973), 127–131.

24. "Review of the Whale Fishery for 1871," *Whalemen's Shipping List and Merchants' Transcript*, February 6, 1872.

25. "Professor Gummere's Speech," in *Old Dartmouth Historical Sketches*, no. 45 (1916), accessed on October 22, 2019, https://www.whalingmuseum.org/explore/library/publications/old-dartmouth-historical-sketches/odhs-no-45.

26. "Review of the Whale Fishery," *Whalemen's Shipping List and Merchants' Transcript* for: January 11, 1881; January 20, 1885; January 25, 1887; January 28, 1890.

Notes—Chapter Six

27. Matthew Howland to Morris Howland, 27 February, 1880, Howland Family Papers, Correspondence of Matthew Howland, 1840–1844. Box 1, Subseries 2, Folder 5, New Bedford Whaling Museum Research Library.

28. Matthew Howland to Morris Howland, 2 March, 1881, Howland Family Papers, Correspondence of Matthew Howland, 1840–1844. Box 1, Subseries 2, Folder 5, New Bedford Whaling Museum Research Library.

29. Matthew Howland to Morris Howland, 28 September 1881, Howland Family Papers, Correspondence of Matthew Howland, 1840–1844. Box 1, Subseries 2, Folder 5, New Bedford Whaling Museum Research Library.

30. Matthew Howland to Morris Howland, 21 November, 1883, Howland Family Papers, Correspondence of Matthew Howland, 1840–1844. Box 1, Subseries 2, Folder 5, New Bedford Whaling Museum Research Library.

31. Pease, *History of New Bedford*, 204; Grieve, *New Bedford Semi-Centennial Souvenir*, 21; Clifford Ashley, *The Yankee Whaler*, reprint of 1926 edition (New York: Dover Publications, 1991), 119.

32. "Edmund Wood," in *Old Dartmouth Historical Sketches*, no. 45 (1916), accessed on October 22, 2019, https://www.whalingmuseum.org/explore/library/publications/old-dartmouth-historical-sketches/odhs-no-45.

33. Zephaniah Pease, *The Centenary of the Merchants National Bank*, 63. It appears Pease lifted the quote from Elbert Hubbard's biography of Robert Owen in Hubbard's "Little journeys to the Homes of Great Business Men" series for the Roycrofters.

34. Peggi Medeiros, "Review: Unitarian minister examines predecessor's life," *South Coast Today*, March 19, 2016, http://www.southcoasttoday.com/entertainmentlife/20160319/review-unitarian-minister-examines-predecessors-life. See also Muderink, *New Bedford's Civil War*, 138, 179; For more on the scope of Potter's congregation see: Richard Allen Kellaway, *William James Potter from Convinced Quaker to Prophet of Free Religion, Vol II* (self-pub., Xlibris, 2015). The Rotches were such prominent members of the church that the steeple contains a bell presented in memory of William and Clara.

35. Creighton, *Rites and Passages*, 77.

36. "Advertisement," *The Ladies' Home Journal*, June 1921, 81.

37. Quoted in Nichols, *Oil & Ice*, 271.

38. Christine Arato and Patrick L. Eleey, *Cultural Landscape Report for New Bedford Whaling National Historical Park* (Boston: National Park Service, 1998).

39. Ashely, *The Yankee Whaler*, 119.

40. Quote reprinted in Rev. Elias Nason, *A Gazetteer of the State of Massachusetts* (Boston: B. B. Russell, 1874), 363.

41. As quoted in: Thomas A. McMullin, "Overseeing the Poor: Industrialization and Public Relief in New Bedford, 1865–1900," *Social Service Review* 65, no. 4 (1991): 555.

42. James M. Lindgren, "'Let Us Idealize Old Types of Manhood': The New Bedford Whaling Museum, 1903–1941," *The New England Quarterly* 72, no. 2 (1999): 170.

43. Ellis, *History of New Bedford*, 557.

44. Rachel Howland to Morris Howland, October 23, 1884, Howland Family Papers, Correspondence of Matthew Howland, 1840–1844. Box 1, Subseries 2, Folder 5, New Bedford Whaling Museum Research Library. Although the archival collection is named primarily for Matthew Howland's correspondence, some letters from Rachel to Morris are included.

45. Thomas Austin McMullin, "Lost Alternative: The Urban Industrial Utopia of William D. Howland," *The New England Quarterly* 55, no. 1 (1982): 32–33.

46. Matthew Howland to Morris Howland, March 11, 1884. Howland Family Papers, Correspondence of Matthew Howland, 1840–1844. Box 1, Subseries 2, Folder 5, New Bedford Whaling Museum Research Library.

47. "The Minors in New Bedford who Cannot Read or Write English," *The Boston Daily Globe*, October 4, 1887.

48. "New Bedford: The Change in its Industrial History," *The Boston Daily Globe*, August 19, 1888.

49. Zephaniah Pease, "Happy Homes of New Bedford's Mill Operatives," *The Boston Daily Globe*, March 16, 1890.

50. "New Bedford City of Whales," Unattributed article in Old Dartmouth Historical Society Scrapbook, #2, 204, New Bedford Whaling Museum Research Library. The article is likely 1901 or 1902 based on its

Notes—Chapter Seven

mention of Captain Jenkin's return from the wreck of the *Kathleen*, which occurred in 1901.

51. "New Bedford's Past," *Daily Mercury*, June 1, 1926.

52. Ashley, *The Yankee Whaler*, 119.

Chapter Seven

1. As quoted in "Chicago's Triumph," *Chicago Daily Tribune*, April 26, 1890.

2. "Sudden Death of Henry E. Weaver," *The Black Diamond*, December 23, 1905, 29; *Industrial Chicago—The Commercial Interests*, Vol. 4 (Chicago: The Goodspeed Publishing Company, 1894), 495-496; "Henry E. Weaver," *The Coal Trade Journal*, December 20, 1905, 920; *Report of the Commissioner of Public Works*, City of Chicago, March 31, 1893, 136; *Official Proceedings, Board of Commissioners*, Cook County Illinois, July 24, 1892, 629-631; "The Coal Trade," *Daily Inter Ocean* [Chicago], January 1, 1892.

3. "Fuel for its Power," *Chicago Daily Tribune*, March 20, 1892; "Fuel for the Fair," *Chicago Daily Tribune*, March 27, 1892; "Electricity at the World's Fair—the Supply of Fuel," *The Electrical World*, December 24, 1892, 405.

4. James Temple Brown, *The Whale Fishery and Its Appliances* (Washington, D.C.: Government Printing Office, 1883).

5. "Exhibit at Louisville," *Annual Report of the Board of Regents of the Smithsonian Institution for the Year Ending June 30, 1886*, Part Two (Washington, DC: Government Printing Office, 1889), 71; "The National Museum Exhibit," *The Courier-Journal* [Louisville, KY], August 26, 1884.

6. Hubert Howe Bancroft, *Book of the Fair* (Chicago: The Bancroft Co., 1893), 518-519.

7. "Whaling City at the World's Fair," *Republican Standard*, April 27, 1893.

8. "A Whale at the Fair," *Chicago Daily Tribune*, August 15, 1892. Reprinted from *New York Tribune*.

9. Jim Coogan, "A Whale, a Tale and the 1893 Chicago Fair," *The Barnstable Patriot*, June 25, 2010. "Fisheries Building," in *Rand, McNally, and Co's Advance Guide to the World Columbian Exposition* (New York: Rand, McNally, & Co, 1893), 60.

10. "A Whaler at the Fair," *The North American* [Philadelphia], November 17, 1891; Reprints with same title: *The Atchison Daily Globe*, November 28, 1891; *Bismarck Daily Tribune*, December 02, 1891.

11. "An Old Whaling Schooner," *Boston Daily Advertiser*, August 21, 1891.

12. "An Act for the Relief of Henry Clay and Others," *The Statutes at Large of the United States of America from December 1889 to March 1891* (Washington, DC: Government Printing Office, 1891), 1366.

13. "Going to the World's Fair," *Republican Standard*, October 8, 1891.

14. "A Whaler at the Fair," *The North American*.

15. "Novel World's Fair Suggestion," *New York Times*, January 12, 1892.

16. "To Chicago Whaling," *Daily Mercury*.

17. "George Bartlett," in *Our County and Its People: Descriptive and Biographical Record of Bristol County, Massachusetts* (Boston: The Boston History Company, 1899), 16; "Remarks by Mr. Headley." in *Old Dartmouth Historical Sketches*, no. 38 (1913), accessed on October 22, 2019, https://www.whalingmuseum.org/explore/publications/old-dartmouth-historical-sketches/odhs-no-38.

18. "The World's Fair," *Republican Standard*, January 14, 1892; "World's Fair Committee," *Daily Mercury*, January 12, 1892.

19. "Progress, bark," *Whalemen's Shipping List and Merchants' Transcript*, May 10, 1892.

20. "For the World's Fair," *Evening Standard*, April 21, 1892.

21. "Members: Union League Club," in *The Chicago Blue Book of Selected Names of Chicago and Suburban Towns* (Chicago: The Chicago Directory Company, 1892), 408-412.

22. "Application for Membership: Henry E Weaver," May 31, 1896, Union League Club Library and Archives; "Committees for 1893," *Director's Record, 1893*, 20-21, Union League Club Library and Archives.

23. Addie Guthrie Weaver, *The Story of Our Flag* (Chicago: A.G. Weaver, 1898), 5.

24. Harriet A. Dunn and Eveline Guthrie Dunn, *Records of the Guthrie Family* (H.A. and S.L Dunn: Chicago, IL, 1898), 81-84; 97.

25. "Union League Members Will Vote by the Australian System," *Chicago Daily Tribune*, January 23, 1892.

26. "W.D. Preston" in *Columbian Expo-*

sition and World's Fair Illustrated (Philadelphia: The Columbia Engraving and Publishing Company, 1893), 175.
27. "Heavy Coal Failure," *Chicago Daily Tribune*, May 29, 1893.
28. "An Important Consideration," *The Chicago Independent*, September, 1890; John Graf, *Chicago's Mansions* (Charleston, SC: Arcadia, 2004), 34.
29. "Arctic Whaling Exhibit Co.," 1892, Corporations, Box 596, No. 25632, Illinois State Archives.
30. "A Whaler at the Fair," *The North American*.
31. "Schooner Franklin," *Evening Standard*, May 18, 1892.
32. "New York and the Big Fair," *New York Times*, May 14, 1892.
33. "The World's Fair," *Republican Standard*.
34. "A Whaling Big Boat," *Daily Inter Ocean*, May 8, 1892.
35. "Bark Progress," *The Evening Standard*, May 18, 1892.
36. Coffin, *"There She Blows,"* 10–11.
37. "To Exhibit an Old Bark," *Chicago Daily Tribune*, May 12, 1892.
38. "New Bedford's Ironsides," *Boston Daily Globe*, May 5, 1892.
39. J.A. Sokalski, *Pictorial Illusionism: The Theatre of Steele MacKaye* (Montreal: McGill-Queen's University Press, 2007), 182.
40. "Note 767," Joseph Anton Solaski, *The Theatre of Steele MacKaye: Pictorial Illusion on the American Stage* (Ph.D. Diss, University of Toronto, 1997), 317; Daniel Leroy Hannon, "The MacKaye Spectatorium," (Ph.D. Diss, Tulane University, 1970), 171.
41. Solaski, *Pictorial Illusionism*, 184.
42. "Bark Progress," *The Evening Standard*.
43. As quoted in: Chris Gaylord, "Herman Melville Books: At first, 'Moby Dick' was a Total Flop," *Christian Science Monitor*, October 18, 2012.
44. "Veteran New England Ships," *Boston Daily Globe*, June 13, 1893.

Chapter Eight

1. Untitled article, *The Ogdensburg Advance and St. Lawrence Weekly Democrat*, July 14, 1892.
2. "Arrived," *Republican Standard*, July 30, 1868.
3. "Monday, June 1, 1868," *Logbook of the Bark Spartan, 1865-1868*, New Bedford Whaling Museum Research Library.
4. Ferguson, *Harpooner*, 26; 237.
5. *Ibid.*, 236.
6. Butler, *J. & W. R. Wing of New Bedford*, 126.
7. "Fifty Years of 'Progress,'" *Montreal Daily Witness*, June 18, 1892.
8. "In a Northeaster Off Cape Sable," *The Evening Standard*, June 23, 1892; "A Chicago Exhibit," *The Quebec Daily Mercury*, June 17, 1892.
9. "To Chicago Whaling," *Daily Mercury*.
10. "Fifty Years of 'Progress,'" *Montreal Daily Witness*.
11. "A Whaler at the World's Fair," *Harper's Weekly*, August 13, 1892, 789; "To Chicago Whaling," *Daily Mercury*.
12. "Destructive Fire in the Woods," *Boston Globe*, April 15, 1892.
13. "Ezra C. Crane," 1880 United States Federal Census, Scio Village, New York, 9; "Ezra Crane," 1900 United States Federal Census, Watkins Village, New York, 3444. "E.C. Crane," 1875 New York State Census; "Cayuga County, New York," *Appointments of U.S. Postmasters, 1832-1971*, 45–46.
14. "E. F. Crane Prosecuted by the Humane Society—He Denies the Charge," *Chicago Tribune*, January 15, 1893; "Ezra F. Crane," *The Lakeside Annual Directory of the City of Chicago*, 1896 (Chicago: The Chicago Directory Company, 1896), 480; Jack Klasey, "The Wealthiest Liveryman in Chicago came from Kankakee," *Daily Journal* [Kankakee, IL], October 1, 2016.
15. For an illustration of E. Fitch Crane see "To Chicago Whaling," *Daily Mercury*.
16. "Fifty Years of 'Progress,'" *Montreal Daily Witness*.
17. *Ibid.*
18. "Marine Matters," *Montreal Herald*, June 17, 1892; "Marine Matters," *Montreal Herald*, June 20, 1892.
19. Coffin, *"There She Blows,"* 42–43.
20. "Fate of the Bark Progress Warning to Those Who Urge Taking Constitution to Chicago," *Evening Standard*, October 26, 1930.
21. "The Famous Whaler Progress Arrives at Ogdensburg," *Daily Journal* [Odgensberg], June 25, 1892.

22. Untitled article, *Daily Journal* [Ogdensburg], June 10, 1892.
23. Untitled article, *Daily Journal* [Odgensburg], June 27, 1892.
24. "A Whaling Big Boat," *Daily Inter Ocean*.
25. Untitled article, *On the St. Lawrence*, June 22, 1892.
26. "Marine News," *Daily Journal* [Ogdensburg], July 1, 1892.
27. "An Old-Time Whaler," *Pittsburg Dispatch*, June 7, 1892; "An Old Whaler," *The Dalles Daily Chronicle*, July 8, 1892; "About the Lakes," *Chicago Daily Tribune*, July 6, 1892; Coffin, "There She Blows," 43.
28. "From the Salt Seas," *Illustrated Buffalo Express*, July 10, 1892.
29. Ibid.
30. "A Whaling Barque," *Detroit Evening News*, July 7, 1892.
31. Dolin, *Leviathan*, 257.
32. The heavily syndicated article appeared in numerous newspapers. All quotes taken from a representative example: "A Whaler at the Fair," *Iowa City Press-Citizen*, August 19, 1892.
33. "A Whaling Barque," *Detroit Evening News*.
34. Miles Orvell, *The Real Thing: Imitation and Authenticity in American Culture, 1880-1940* (Chapel Hill: University of North Carolina Press, 1989), xvii.
35. Untitled article, *The Daily Journal* [Ogdensburg], June 22, 1892.
36. "Lake Marine News," *Daily Inter Ocean*, July 15, 1892; "News of the Lakes," *Milwaukee Journal*, July 14, 1892.
37. "Old Whaling Ship at Racine," *Milwaukee Sentinel*, July 15, 1892.
38. "Progress in Port," *Journal Times* [Racine], July 14, 1892.
39. "A Great Attraction," *Journal Times*, July 15, 1892.
40. "Whaler Progress," *Journal Times*, July 21, 1892.
41. "Saturday July 23, 1892," *Diary of Captain Soren Kristiansen, Lake Michigan Schooner Captain—1891-1893*, Delta County Historical Society, accessed on October 22, 2019 http://www.uproc.lib.mi.us/SecordPress/DiarySoren/Kristiansen2.htm.
42. "Almost a Mutiny Resulted," *Milwaukee Sentinel*, July 22, 1892.
43. "The Progress in Trouble," *Journal Times*, July 22, 1892.
44. Briton Cooper Busch, *Whaling Will Never Do for Me* (Lexington: University of Kentucky Press, 1994), 79.
45. "The Progress in Trouble," *Journal Times*; "Out of a Job," *The Journal Times*, July 23, 1892.
46. "Whaler Progress," *Journal Times*; The Arctic Whaler Progress A Complete Marine and Whaling Museum.
47. Allen, *Children of the Light*, 183.
48. Amy Koritz, "Re/Moving Boundaries: From Dance History to Cultural Studies," in *Moving Words: Re-Writing Dance*. ed. Gay Morris (London: Routledge, 2002), 83.
49. "Fate of the Bark Progress Warning to Those Who Urge Taking Constitution to Chicago," *The Evening Standard*.
50. "Carter All Right," *Journal Times*, July 26, 1892. Carter would later recount his experience in 1930, including describing the *Progress* as perennially leaking and rotten, a narrative that doesn't entirely match others. Given his negative association with the enterprise—and the amount of time that had passed—only a few quotes from the article have been used here, rather than treating it as a major primary source. See "Fate of the Bark Progress Warning to Those Who Urge Taking Constitution to Chicago," *Evening Standard*.
51. Progress in Port," *Journal Times*.

Chapter Nine

1. Solaski, *The Theatre of Steele MacKaye*; Hannon. *The MacKaye Spectatorium*, 12.
2. "Good Ship Progress," *Chicago Herald*, July 28, 1892. Reprinted as "Progress in Port," *New Bedford Daily Mercury*, August 1, 1892.
3. "Good Ship Progress," *Chicago Herald*.
4. Ibid.; Additional accounts from July 27 are: "Arctic Sea Trophies," *Daily Inter Ocean*, July 28, 1892; "Bark Progress Here," *Chicago Tribune*, July 28, 1892.
5. "To Take Charge of the Progress," *Republican Standard*, June 16, 1892.
6. "An Ancient Mariner," *Daily Inter Ocean*, July 19, 1892.
7. Selected quotes in order from: "Good Ship Progress," *Chicago Herald*; "An Ancient

Notes—Chapter Nine

Mariner," *Daily Inter Ocean*; "Bark Progress Here," *Chicago Tribune*; "Arctic Sea Trophies," *Daily Inter Ocean*.

8. "Currents of Commerce," *Boston Daily Globe*, November 26, 1892.

9. Daniel P. Toomey and Thomas Quinn, eds., *Massachusetts of To-Day* (Boston: Columbia Publishing Company, 1892), 356.

10. Pease, *New Bedford: Its History, Industries, Institutions and Attractions*, 35.

11. "Whales at the Fair," *Chicago Daily Tribune*, July 24, 1892. All subsequent quotes describing a tour around the *Progress* from the same.

12. "Saturday July 23, 1892," *Diary of Captain Soren Kristiansen*.

13. Selected quotes in order from: "New York and the Big Fair," *New York Times*; "Novel World's Fair Suggestion," *New York Times*, January 12, 1892; "The Famous Whaler Progress Arrives in Ogdensburg," *The Daily Journal* [Odgednsburg]. The last also includes the "polar bear skins" mention.

14. "Good Ship Progress," *Chicago Herald*; "Arctic Sea Trophies," *Daily Inter Ocean*.

15. "Fanned by Arctic Breezes," *Chicago Times*, July 28, 1892.

16. "MacKay [sic], James Steele, The Mystery of Emotion," *Index of Speakers at New Bedford Lyceum, 1855–1907*. MSS 7, Seq 1, Series F, 260. New Bedford Whaling Museum Research Library.

17. Sokalski, *Pictorial Illusionism*, 183.

18. "Advertisement," *Chicago Daily Tribune*, July 30, 1892; "Advertisement," *Chicago Daily Tribune*, August 19, 1892.

19. "For the World's Fair," *Emporia Daily Gazette*, June 17, 1892.

20. "Bark Progress Here," *Chicago Tribune*.

21. Untitled article, *Milwaukee Journal*, July 28, 1892.

22. "Advertisement," *Chicago Daily Tribune*, August 5, 1892.

23. "The Whaler Progress," *Chicago Mail*, August 6, 1892.

24. "Old Progress Open to the Public," *Chicago Daily Tribune*, August 6, 1892.

25. As quoted in Busch, *Whaling Will Never Do for Me*, 187.

26. The World's Fair," *Republican Standard*.

27. "Old Progress Open to the Public," *Chicago Daily Tribune*.

28. James W. Cook, *The Arts of Deception* (Cambridge: Harvard University Press, 2001).

29. Hannon, *The MacKaye Spectatorium*, 54.

30. The Arctic Whaler Progress A Complete Marine and Whaling Museum.

31. Brown, *The Whale Fishery and Its Appliances*, 6; 17.

32. D. Graham Burnett, *Trying Leviathan* (Princeton: Princeton University Press, 2007), 39.

33. Phineas Taylor Barnum, *The Autobiography of P.T. Barnum* (London: Ward and Lock, 1855), 85.

34. "Bark Progress," *Republican Standard*, August 11, 1892.

35. "Voyage Ended," *Republican Standard*, August 4, 1892.

36. The tiles eventually ended up in the collection of the Peabody Essex Museum in Salem, MA. ID M876.AB with object note: "The tiles may have been produced as souvenir's [sic] from the "Arctic Whaling Exhibition" at the 1893 Chicago World's Fair." Bliss & Nye produced similar souvenir plates, mugs and pitchers for other fairs and civic anniversaries.

37. "Bark Progress," *Republican Standard*, August 11, 1892.

38. "Bark Palmetto Sold," *Daily Mercury*, September 15, 1892.

39. Ibid.; "Whalers on the Lakes," *Republican Standard*, September 15, 1892. Syndicated examples of the 25-word story include "News of the Day," *Weekly Herald* (Shenandoah, Pennsylvania), September 17, 1892.

40. "Seeking Whales in New Seas," *New York Times*, August 8, 1893.

41. Ibid.

42. "Mrs. Sherwood's Reading," *Evening Standard*, undated article in *William Bradford Scrapbook 1880–1892*, New Bedford Whaling Museum Research Library.

43. Ellis, *History of New Bedford and Its Vicinity*, 454; 434.

44. Dolan, *Leviathan*, 115–116.

45. "Bay State Whaling," *Sacramento Daily Union Record*, January 30, 1898.

46. "The Whaling Bark Progress," *Chicago Evening Journal*, August 4, 1892.

47. "Exit the Progress," *Chicago Daily Tribune*, September 25, 1892.

48. Sokalski, *Pictorial Illusionism*, 230.

Notes—Chapter Nine

49. "The Whaler Progress at Chicago," *Brooklyn Daily Eagle*, July 28, 1892.

50. "Exit the Progress," *Chicago Daily Tribune*; the alternative name of Albert found in "Wrecked in Port," *Daily Inter Ocean*, September 25, 1892; On Dowden's return: "Personal," *Daily Mercury*, September 16, 1892.

51. "Exit the Progress," *Chicago Daily Tribune*.

52. Ibid.

53. *The Arctic Whaler Progress A Complete Marine and Whaling Museum*.

54. "Exit the Progress," *Chicago Daily Tribune*; "Wrecked in Port," *Daily Inter Ocean*; "Raising the Old Whaler," *Chicago Daily Tribune*, September 26, 1892.

55. "Exit the Progress," *Chicago Daily Tribune*; "Wrecked in Port," *Daily Inter Ocean*.

56. "Whaler Progress Sunk," *Milwaukee Sentinel*, September 25, 1892.

57. "Exit the Progress," *Chicago Daily Tribune*; "Wrecked in Port," *Daily Inter Ocean*.

58. "Whaler Progress Sunk," *Milwaukee Sentinel*; Arctic Whaling Exhibit Company vs. Tug James Hay, 1892 District Court of the United States for the District of In Admiralty, Record Group 21, Admiralty Records, National Archives, Chicago.

59. "Exit the Progress," *Chicago Daily Tribune*.

60. "Sunk the Progress," *The Chicago Evening Post*, October 22, 1892.

61. "Wrecked in Port," *Daily Inter Ocean*.

62. "Exit the Progress," *Chicago Daily Tribune*.

63. "Wrecked in Port," *Daily Inter Ocean*.

64. Arctic Whaling Exhibit Company vs. Tug James Hay.

65. "Exit the Progress," *Chicago Daily Tribune*; "Wrecked in Port," *Daily Inter Ocean*. The only change in text between the two was the *Inter Ocean's* use of female pronouns.

66. "Sunk the Progress," *The Chicago Evening Post*.

67. "Whaler Progress Sunk," *Milwaukee Sentinel*.

68. "Many Children in Peril," *Chicago Journal*, September 24, 1892.

69. Arctic Whaling Exhibit Company vs. Tug James Hay.

70. "Exit the Progress," *Chicago Daily Tribune*.

71. "Wrecked in Port," *Daily Inter Ocean*.

72. "Was Sunk by a Sand Scow," *Chicago Times*, September 25, 1892.

73. "Raising the Old Whaler," *Chicago Daily Tribune*, September 26, 1892.

74. "Boarded by Pirates," *Chicago Daily Tribune*, September 27, 1892.

75. "Raising the Old Whaler," *Chicago Daily Tribune*.

76. See as representative example: "Accident to the Whaler Progress," *New York Times*, September 25, 1892.

77. "Raising the Old Whaler," *Chicago Daily Tribune*; "Raising the Sunken Whaler," *Chicago Record*, September 26, 1892.

78. "The Old Whaling Bark 'Progress,'" *Chicago Daily Tribune*, October 15, 1892. For the actual date of towing: "Exposition Notes," *Chicago Daily Tribune*, November 27, 1892.

79. "World's Fair Notes," *Chicago Daily Tribune*, December 24, 1892; "Construction Goes on as Usual," *Chicago Daily Tribune*, January 7, 1893.

80. F. Penn, "Newspaper Artists—G. A. Coffin," *The Inland Printer*, June, 1894, 241.

81. Coffin, *"There She Blows,"* 5; 8.

82. Ibid., 46.

83. Ibid., 10; 47

84. *Report of the President to the Board of Directors of the World's Columbian Exposition* (Chicago: Rand, McNally, and Co., 1898), 93; "Information of Importance to the Public" [promotional brochure], April 10, 1893, Reel 1, Box 25, Folder 12, Moses P. Handy Papers, Chicago Public Library, Special Collections.

85. *Conkley's Complete Guide to the World's Columbian Exposition* (Chicago: W. B. Conkley Company, 1893), 25.

86. *Rand, McNally, & Co's A Week at the Fair* (Chicago: Rand, McNally, and Co., 1893), 111.

87. Julian Ralph, *Harper's Chicago and World's Fair* (New York: Harper & Brothers Publishers, 1893), 243.

88. Trumbull White and William Igleheart, *The World's Columbian Exposition, Chicago, 1893* (Philadelphia: International Publishing Co., 1893), 229.

89. "In the Back Yard," *World's Fair Puck*, July 17, 1893.

90. "Whaler Ditched," *Boston Daily Advertiser*, June 30, 1896.

91. Tudor Jenks, *The Century World's Fair*

Notes—Chapter Ten

Book for Boys and Girls (New York: The Century Company, 1893), 47–49.
92. Mary L. Sherman, "The World's Fair," *Los Angeles Times*, September 11, 1893.
93. Mrs. Mark Stevens, *Six Months at the World's Fair* (Detroit: Detroit Free Press Printing Company, 1895), 20–21.
94. "Special World's Fair Letter," *The Poultney Journal* [Poultney, Vermont], June 30, 1893.
95. "Veteran New England Ships," *Boston Daily Globe*, June 3, 1893.
96. "Heavy Coal Failure," *Chicago Daily Tribune*.
97. "The Weaver Letter," *Director's Record*, 1893, 61–62, Union League Club Library and Archives.
98. "Reception and War Dances," *Chicago Daily Tribune*, August 18, 1893.
99. Sherman, "The World's Fair."
100. "Fishermen's Days," *Boston Daily Globe*, September 1, 1893.
101. "August, September, and October," in *Report of the President to the Board of Directors of the Columbian Exposition*, 257–258. Also, "Fishermen Begin Swapping Yarns," *Chicago Daily Tribune*, September 20, 1893; "A Review of the General Condition of the World Fair Buildings," *Brooklyn Daily Eagle*, April 30, 1893.
102. "Weaver-Getz Company Reorganized," *Chicago Daily Tribune*, October 18, 1893.
103. "Fate of the Bark Progress Warning to Those Who Urge Taking Constitution to Chicago," *The Evening Standard*.
104. "Going to the World's Fair," *Republican Standard*.
105. Sokalski. *Pictorial Illusionism*, 186.
106. "Being Demolished," *Republican Standard*, October 19, 1893.

Chapter Ten

1. As quoted in: Marla R. Miller and Laura A. Miller, *A Generous Sea: Native Hawaiians, Pacific Islanders, and the Jewish Community in New Bedford Whaling & Whaling Heritage* (Boston: Northeast Region Ethnography Program, National Park Service, 2016), 201.
2. Details from Howland as quoted in Lindgren, "Let Us Idealize Old Types of Manhood," 171–172.
3. Miller, *A Generous Sea*, 197–198.
4. "Historical Society Advised," unattributed newspaper clipping, January 17, 1908 Old Dartmouth Historical Society Scrapbook, #1, 1, New Bedford Whaling Museum Research Library.
5. Wood quote: "Proceedings of the Annual Meeting of the Old Dartmouth Historical Society, New Bedford, Massachusetts, on March 31, 1904" in *Old Dartmouth Historical Society Sketches* no. 5 (1904), accessed on October 22, 2019 https://www.whalingmuseum.org/explore/library/publications/old-dartmouth-historical-sketches/odhs_no_5.
6. Proceedings of the Annual Meeting…1904"; Lindgren, "Let Us Idealize Old Types of Manhood," 170; "Historical Society Advised."
7. "Whaling Exhibit is Purchased," *Chicago Daily Tribune*, February 21, 1894.
8. "Fate of the Whaler Progress," *Chicago Daily Tribune*, July 22, 1895.
9. "Fate of the Whaler Progress," *Washington Post*, July 4, 1896. Reprinted from *Chicago Chronicle*.
10. "Neglected Caravels Going to Pieces," unattributed newspaper clipping, series H, subseries 2, folder 1, Nicholson Whaling Manuscript Collection, Providence Public Library.
11. "I.H. Bartlett & Sons," *Deposits*, May 11, 1892, The Records of the Merchants Bank/Merchants National Bank of New Bedford, Massachusetts, New Bedford Whaling Museum Research Library; "Income from Ships to 9th Month, 28, 1891," J. & W. R. Wing & Company Records, 1833–1918, box 16, series C, subseries 1, folder 1, New Bedford Whaling Museum Research Library.
12. *Arctic Whaling Exhibit Company vs. Tug James Hay*.
13. "Arctic Whaling Exhibit Co.," Illinois State Archives.
14. *Arctic Whaling Exhibit Company vs. Tug James Hay*.
15. "New Incorporations," *Chicago Daily Tribune*. July 21, 1891; "New Illinois Corporations," *The Daily Inter Ocean*, July 21, 1891.
16. "Whaling Exhibit is Purchased," *Chicago Daily Tribune*; The *Arctic Whaler Progress*.
17. "Section of Fisheries," in *Guide to the Field Columbian Museum: With Diagrams*

183

Notes—Chapter Ten

and Descriptions, First Edition (Chicago: Field Columbian Museum, 1894), 171.

18. Ibid.

19. "A Whaler at the World's Fair," *Harper's Weekly*.

20. "Section of Fisheries." 171–172.

21. "Alterations in Installation," front piece note in *Guide to the Field Columbian Museum: With Diagrams and Descriptions*. Third Edition. (Chicago: Field Columbian Museum, 1895).

22. "Annual Report of the Director," in *Annual Report of the Director to the Board of Trustees for the Year 1895-96* (Chicago: Field Columbian Museum, 1896), 87.

23. "Whaler Progress Material" [notes from various departments], November 6, 1907, Memo #1742 Peabody Institute, Field Museum Library and Archives; "Accession Record," January 21, 1895, Historical Record 24, Department Z, Field Museum Library and Archives.

24. Paul D. Brinkman, "'The 'Chicago Idea': Patronage, Authority, and Scientific Autonomy at the Field Columbian Museum, 1893-97," *Museum History Journal* 8, no. 2 (2015): 168-187; Robert R. Kargon, "The Civic Culture of Modernity in Chicago, 1880-1910," in *Urban Modernity: Cultural Innovation in the Second Industrial Revolution* (Cambridge: The MIT Press, 2010), 150. For examples of minor whaling artifact displays in or off the West Court see the Fourth and Fifth editions of *Guide to the Field Columbian Museum: With Diagrams and Descriptions*, 1896; 1897.

25. Dr. Dorsey, Curator of Anthropology to Director Skiff, 12 June, 1907, Memo #1742 Peabody Institute, Field Museum Library and Archives.

26. Director Skiff to Prof. E. S. Morse, Peabody Museum., 19 June, 1907. Memo #1742 Peabody Institute, Field Museum Library and Archives.

27. "Material from the Whaler 'Progress'—Five Boxes and Four Crates," undated; "Invoice from Michigan Central Railroad Company," October 2, 1907; Director Skiff to Prof. E. S. Morse, Peabody Museum, 25 September, 1907; all located in Memo #1742 Peabody Institute, Field Museum Library and Archives.

28. "Fate of the Whaler Progress," *Chicago Daily Tribune*. The article only mentions a "man named Peterson." Later articles would add "J.K. Peterson."

29. Untitled article, *Chicago Tribune*, December 27, 1894.

30. "Fate of the Whaler Progress," *Chicago Daily Tribune*.

31. Untitled article, *Chicago Tribune*, June 26, 1896.

32. "Fate of the Whaler Progress," *Washington Post*.

33. "New Use for Whaler Progress," *Chicago Daily Tribune*, July 11, 1896; Also, Josiah Seymour Currey, "Peter A. Newton." in *Chicago: Its History and its Builders, Volume 5* (Chicago: S.J. Clark Publishing Company, 1912), 105–106.

34. "Whaler Coming Back from Chicago," *New York Times*, July 13, 1896; "Raise a Sunken Vessel," *Chicago Record*, July 11, 1896.

35. "Old Whaler is Removed," *Daily Inter Ocean*, July 11, 1896; Untitled article, *Chicago Journal*, July 11, 1896.

36. Untitled article, *Chicago Daily Tribune*, July 20, 1896.

37. "Boats and Yachts," *Chicago Daily Tribune*, June 12, 1897; "Boats and Yachts," *Chicago Daily Tribune*, June 13, 1897.

38. "Dynamite for an Old Whaler," *Chicago Daily Tribune*; "Whaler Progress, World's Fair Relic, Soon to be Demolished," *Chicago Daily Tribune*, December 22, 1901.

39. "Whaler is Lost in Sea of Fire," *Chicago American*, February 26, 1892. "World's Fair Whaler is made a Wreck by Fire," *Chicago Daily Tribune*, February 26, 1902.

40. "The Progress as a Bark," *Westerly Sun*, January 17, 1902.

41. "Fate of the Whaler Progress," *Washington Post*.

42. "At the White House," *Evening Star* [Washington D.C.], October 11, 1897.

43. Untitled article, *Buffalo Sunday Morning News*, March 2, 1902.

44. "The Fate of the Progress," *Evening Standard*; "The Old Progress' Fate," *Evening Standard*, June 30, 1896; "The Progress' Last Days," *Evening Standard*, June 9, 1899.

45. "The Progress' Last Days," *Evening Standard*; "Old Whaler Progress," *Westerly Sun*, July 12, 1899.

46. Lindgren, "Let Us Idealize Old Types of Manhood," 177.

47. *Old Dartmouth Historical Sketches*,

Notes—Chapter Ten

no. 38 (1913), accessed on October 22, 2019, https://www.whalingmuseum.org/explore/library/publications/old-dartmouth-historical-sketches/odhs-no-38

48. "Alanson Borden," in *Our Country and Its People: A Descriptive and Biographical Record of Bristol County, Massachusetts, Part 2* (Boston: The Boston History Company, 1899), 16.

49. "New Bedford Wharves Once Again Alive with Whalermen," *The Scranton Republican*, September 15, 1905.

50. "Sudden Death of Henry Weaver," *The Black Diamond*.

51. "Whales not in Demand," *The Daily Republican* [Monongahela, Pennsylvania], September 12, 1902.

52. "E. F. Crane Prosecuted by the Humane Society—He Denies the Charge." *Chicago Tribune*; "Ezra F. Crane," *The Lakeside Annual Directory*; "Father-in-Law of Olean Man had Rare Relic," *Times Herald* [Olean, NY], September 5, 1925; "Entertains his Friends at Dinner at Breesport," *Star-Gazette* [Elmira, NY], November 15, 1913; "Ezra F. Crane," *Democrat and Chronicle* [Rochester, NY], March 8, 1927.

53. "Capt. Gifford Caught Fifteen Whales," *New York Times*, April 21, 1893; "Obituary," *Daily Mercury Standard*, December 11, 1899.

54. "Captain Dowden and Bark Progress," *The Leavenworth Times*, February 4, 1908.

55. "Hero of Arctic Pack of 1871 Dead," *Boston Daily Globe*.

56. "New Yorker Seeks Old Whaler," *New York Times*, March 9, 1939.

57. "Exhibiting Old Whaler at N.Y. Fair Opposed," *Christian Science Monitor*, March 10, 1939.

58. Everett S. Allen, *1941 Story Recounts Journey to Mystic*, June 20, 2014, http://www.southcoasttoday.com/article/20140620/news/406200303.

59. Ibid.

Bibliography

Archives and Collections

Admiralty Records (1871–1942), National Archives, Chicago.
City Documents, New Bedford Free Public Library.
David Batchelder Kempton Papers, New Bedford Whaling Museum Research Library.
Howland Family Papers, Correspondence of Matthew Howland, New Bedford Whaling Museum Research Library.
Illinois State Archives.
J. & W. R. Wing & Company Records, 1833–1918, New Bedford Whaling Museum Research Library.
Library and Archives, Field Museum of Natural History, Chicago.
Maritime Art and History Collections, Peabody Essex Museum.
Matthew Howland Letter Book, 1858–1879, Baker Library Special Collections, Harvard University.
Moses P. Handy Papers, Chicago Public Library.
Nicholson Whaling Collection, Providence Public Library.
Old Dartmouth Historical Society Scrapbook Collection, New Bedford Whaling Museum Research Library.
Old Dartmouth Historical Society Sketches, New Bedford Whaling Museum Research Library.
The Records of the Merchants Bank/Merchants National Bank of New Bedford, New Bedford Whaling Museum Research Library.
Union League Club Library and Archives.
Westerly Local History Collection, Westerly Public Library & Wilcox Park.
William Bradford Scrapbooks, New Bedford Whaling Museum Research Library.

Primary Sources: Newspapers and Periodicals

Boston Daily Globe.
Chicago American.
Chicago Daily Tribune.
Chicago Evening Journal.
Chicago Herald.
Chicago Mail.
Chicago Record.
Daily Inter Ocean [Chicago].
Evening Standard [New Bedford].
Journal Times [Racine].
New Bedford Mercury, Daily Mercury, and *Daily Mercury Standard.*
New York Times.
Republican Standard [New Bedford].
Whalemen's Shipping List and Merchants' Transcript.

Additional Primary Sources

Annual Report of the Director to the Board of Trustees for the Year 1895–96. Chicago: Field Columbian Museum, 1896.
The Arctic Whaler Progress A Complete Marine and Whaling Museum. Chicago: Knight, Leonard & Co., 1892.
Bancroft, Hubert Howe. *Book of the Fair.* Chicago: The Bancroft Co., 1893.
Barnum, Phineas Taylor. *The Autobiography of P.T. Barnum.* London: Ward and Lock, 1855.
"Beyond the Arctic Circle." *The Contributor,* January, 1895.
Brown, James Temple. "Stray Leaves from a Whaleman's Log." *Century Illustrated Magazine,* February 1893.
———. *The Whale Fishery and Its Appliances.* Washington, D.C.: Government Printing Office, 1883.

Bibliography

Cheever, Henry T. *The Whale and his Captors*. New York: Harper and Brothers, 1864.

The Chicago Blue Book of Selected Names of Chicago and Suburban Towns. Chicago: The Chicago Directory Company, 1892.

Coffin, G. A. *"There She Blows" or the Story of the Progress*. Chicago: The Arctic Whaling Expedition Company, 1893.

Columbian Exposition and World's Fair Illustrated. Philadelphia: The Columbia Engraving and Publishing Company, 1893.

Conkley's Complete Guide to the World's Columbian Exposition. Chicago: W. B. Conkley Company, 1893.

"Digitized Logbooks." New Bedford Whaling Museum Research Library. Accessed October 22, 2019. https://www.whalingmuseum.org/explore/library/logbooks-journals.

Dunn, Harriet A. and Eveline Guthrie Dunn. *Records of the Guthrie Family*. H.A. and S.L Dunn: Chicago, IL, 1898.

Ellis, Leonard Bolles. *History of New Bedford and Its Vicinity*. Syracuse: D. Mason & Co., 1892.

Ferguson, Robert. *Harpooner*. Philadelphia: University of Pennsylvania Press, 1936.

Grieve, Robert, ed. *New Bedford's Semi-Centennial Souvenir*. Providence: Journal of Commerce Company, 1897.

Guide to the Field Columbian Museum: With Diagrams and Descriptions, First Edition. Chicago: Field Columbian Museum, 1894.

_____. *Fifth Edition*. Chicago: Field Columbian Museum, 1897.

_____. *Fourth Edition*. Chicago: Field Columbian Museum, 1896.

_____. *Third Edition*. Chicago: Field Columbian Museum, 1895.

Industrial Chicago—The Commercial Interests, Vol. 4. Chicago: The Goodspeed Publishing Company, 1894.

Jenks, Tudor. *The Century World's Fair Book for Boys and Girls*. New York: The Century Company, 1893.

Kingston, William Henry Giles. *Peter the Whaler*. New York: E.P. Dutton & Co., 1906.

Kirk, William, ed. *A Modern City: Providence, Rhode Island and Its Activities*. Chicago: University of Chicago Press, 1909.

"Librarian's Report." In *The Fiftieth Annual Report of the Trustees of the Free Public Library of the City of New Bedford*. New Bedford: A.E. Coffin Press, 1902.

Melville, Herman. *Moby-Dick*. Pleasantville, NY: The Reader's Digest Association, 1989.

Morfit, Campbell. *A Treatise on Chemistry Applied to the Manufacture of Soap and Candles*. Philadelphia: Parry and McMillian, 1856.

Nason, Elias, Rev. *A Gazetteer of the State of Massachusetts*. Boston: B. B. Russell, 1874.

"New Bedford, Massachusetts." *The National Magazine and Industrial Record* 1, no. IV (1845): 328–343.

"Occasional Notes." *The Pall Mall Budget*, December 6, 1879.

Olmsted, Frances Allyn. *Incidents of a Whaling Voyage*. New York: D. Appleton and Co., 1841.

Our County and Its People: A Descriptive and Biographical History of Bristol County, Massachusetts. Boston: The Boston History Company, 1899.

Owners and Crew of the Hawaiian Bark Arctic. SR 577. 52rd Cong., 1st sess., Reports of Committees of the United States Senate for the First Session of the Fifty-Second Congress, Vol 4: 1–25.

Peabody, Ephraim. *A Discourse Delivered at the First Public Meeting of the New Bedford Orphan's Home*. Boston: Isaac R. Butts, 1842.

Pease, Zephaniah W. *The Centenary of the Merchants National Bank*. New Bedford: Reynolds Printing, 1925.

_____. *History of New Bedford*. New York: The Lewis Historical Publishing Company, 1918.

_____. *New Bedford, Massachusetts: Its History, Industries, Institutions, and Attractions*. New Bedford: Mercury Publishing Company, 1889.

Penn, F. "Newspaper Artists—G. A. Coffin." *The Inland Printer*, June, 1894.

Ralph, Julian. *Harper's Chicago and World's Fair*. New York: Harper & Brothers Publishers, 1893.

Rand, McNally, and Co's A Week at the Fair. Chicago: Rand, McNally, and Co., 1893.

Rand, McNally, and Co's Advance Guide to the World Columbian Exposition. New York: Rand, McNally, and Co., 1893.

Report of the Officers Constituting the Light-House Board. Washington, D.C.: A. Boyd Hamilton, 1852.

Bibliography

Report of the President to the Board of Directors of the World's Columbian Exposition, Chicago: Rand, McNally, and Co., 1898.
Seward, William Henry. *The Whale Fishery, and American Commerce in the Pacific Ocean. Speech of William H. Seward, in the Senate of the United States, July 29, 1852.* Washington, D.C.: Buell & Blanchard, 1852.
Spahr, Charles B. *America's Working People.* London: Longmans, Green, and Co., 1900.
Starbuck, Roger. *The Golden Harpoon or, Lost Among the Floes: A Story of the Whaling Grounds.* New York: Beadle and Adams, Publishers, 1865.
Stevens, Mark, Mrs. *Six Months at the World's Fair.* Detroit: Detroit Free Press Printing Company, 1895.
Stewart, Sturgeon. "The Whale and his Haunts, Article II." *The Canadian Magazine,* October 1909.
Toomey, Daniel P., and Thomas Quinn, eds. *Massachusetts of To-Day.* Boston: Columbia Publishing Company, 1892.
Tower, Walter Sheldon. *A History of the American Whale Fishery.* Philadelphia: University of Pennsylvania Press, 1907.
Weaver, Addie Guthrie. *The Story of Our Flag.* Chicago: A.G. Weaver, 1898.
White, Trumbull, and William Igleheart. *The World's Columbian Exposition, Chicago, 1893.* Philadelphia: International Publishing Co., 1893.
Williams, Eliza Azelia. "The Voyage of the Florida, 1858–1861: From the Journal of Eliza Azelia Williams." In *One Whaling Family,* edited by Harold Williams, 3–204. Boston: Houghton Mifflin Company, 1964.
Williams, William Fish. "The Voyage of the Florence, 1873–1874." In *One Whaling Family,* ed. Harold Williams, 245–383. Boston: Houghton Mifflin Company, 1964.

Secondary Sources

Allen, Everett S. *Children of the Light.* Boston: Little, Brown, and Company, 1973.
Arato, Christine, and Patrick L. Eleey. *Cultural Landscape Report for New Bedford Whaling National Historical Park.* Boston: National Park Service, 1998.
Ashley, Clifford. *The Yankee Whaler* (reprint of 1926 edition). New York: Dover Publications, 1991.
Bockstoce, John R. *Furs and Frontiers in the Far North: The Contest among Native and Foreign Nations for the Bering Strait Fur Trade.* New Haven: Yale University Press, 2009.
Brickman, Paul D. "'The 'Chicago Idea': Patronage, Authority, and Scientific Autonomy at the Field Columbian Museum, 1893–97." *Museum History Journal* 8, no. 2 (2015): 168–187.
Brown, Ronald R. *Dying on the Job: Murder and Mayhem in the American Workplace.* New York: Rowman and Littlefield, 2013.
Burg, Ryan S. "Rebuilding the Iron Cage: Post-Failure Organizing in Newspapers and Investment Banks." Ph.D. Diss., University of Pennsylvania, 2011.
Burnett, D. Graham. *Trying Leviathan.* Princeton: Princeton University Press, 2007.
Busch, Briton Cooper. *Whaling Will Never Do for Me.* Lexington: University of Kentucky Press, 1994.
Butler, Joseph Martin. "J. & W. R. Wing of New Bedford: A Study of the Impact on a Declining Industry upon an American Whaling Agency." Ph.D. diss., Pennsylvania State University, 1973.
Chippendale, H.A. *Sails and Whales.* Boston: Houghton Mifflin, 1951.
Clayton, Barbara, and Kathleen Whitley. *Guide to New Bedford.* Chester, CT: The Globe Pequot Press, 1979.
Cook, James W. *The Arts of Deception.* Cambridge: Harvard University Press, 2001.
Crabtree, Sarah Lelia. "A Holy Nation: The Quaker Itinerant Ministry in an Age of Revolution, 1750–1820." Ph.D. Diss., University of Minnesota, 2007.
Creighton, Margaret S. *Rites & Passages: The Experience of American Whaling, 1830–1870.* Cambridge: Cambridge University Press, 1995.
Davis, Lance E., Robert E. Gallman, and Karin Gleiter. *In Pursuit of Leviathan: Technology, Institutions, Productivity, and Profits in American Whaling, 1816–1906.* Chicago: University of Chicago Press, 1997.
Dolin, Eric Jay. *Leviathan: The History of Whaling in America.* New York: W.W. Norton, 2007.

Bibliography

Dublin, Thomas, ed. *Farm to Factory: Women's Letters, 1830-1860*. New York: Columbia University Press, 1981.

Edwards, Mary K. Bercaw. "An Old Sailor's Lament: Herman Melville, the Stone Fleet, and the Judgment of History." *Leviathan* 9, no. 3 (2007): 51-64.

Feinman, Peter. "Whaling and Abolition: A Sample 'Path Through History.'" *Institute of History, Archeology and Education*. April 29, 2013, https://ihare.org/2013/04/29/whaling-and-abolition-a-sample-path-through-history/.

Gaines, W. Craig. *Encyclopedia of Civil War Shipwrecks*. Baton Rouge: Louisiana State University Press, 2008.

Graf, John. *Chicago's Mansions*. Charleston, SC: Arcadia, 2004.

Graves, Michael P. "Functions of Key Metaphors in Early Quaker Sermons, 1671-1700." *Quarterly Journal of Speech* 69, no. 4 (1983): 364-378.

Greene, Robert A. "Whichcote, the Candle of the Lord, and Synderesis." *Journal of the History of Ideas* 52, no. 4. (1991): 617-644.

Haley, Nelson Cole. *Whale Hunt*. New York: Ives Washburn, Inc., 1948.

Hannon, Daniel Leroy. "The MacKaye Spectatorium." Ph.D. Diss, Tulane University, 1970.

Heath, Kingston Wm. *The Patina of Place: The Cultural Weathering of a New England Industrial Landscape*. Knoxville: University of Tennessee Press, 2001.

Hintze, Catherine M., and Robert Mulero. *The History and Design of New York City Streetlights, Past and Present*. Pittsburgh: Dorrance Publishing Co., 2017.

"Historic Warm White Street Lights in Washington." *Capitol Hill Restoration Society*. March, 2017. http://chrs.org/wp-content/uploads/2017/03/StreetLghtsWashingtonHistory4-1-2017eap.pdf.

Hoare, Philip. *The Whale: In Search of the Giants of the Sea*. New York: ECCO, 2010.

Hohman, Elmo P.. *The American Whaleman* (reprint of 1928 edition). Clifton: Augustus M Kelley, 1972.

"Interpretive Themes: Unique Cultures." New Bedford Whaling National Historical Park. Accessed April 15, 2019. https://www.nps.gov/nebe/learn/historyculture/cultures.htm.

Jaffee, David. *A New Nation of Goods: The Material Culture of Early America*. Philadelphia: University of Pennsylvania Press, 2010.

Kargon, Robert R. "The Civic Culture of Modernity in Chicago, 1880-1910." In *Urban Modernity: Cultural Innovation in the Second Industrial Revolution*. Cambridge: The MIT Press, 2010.

Kellaway, Richard Allen. *William James Potter from Convinced Quaker to Prophet of Free Religion, Vol II*. N.p.: Xlibris, 2015.

Kirschenblatt-Gimblett, Barbara. *Destination Culture: Tourism, Museums, and Heritage*. Berkeley: University of California Press, 1998.

Koritz, Amy. "Re/Moving Boundaries: From Dance History to Cultural Studies." In *Moving Words: Re-Writing Dance*. Ed. Gay Morris. London: Routledge, 2002.

Laborde, Peggy Scott, and John Magill. *Canal Street: New Orleans' Great Wide Way*. Gretna, LA: Pelican Publishing Company, 2006.

Larson, Erik. *The Devil in the White City*. New York: Crown Publishers, 2003.

Lewis, Arnold. *An Early Encounter with Tomorrow*. Urbana: University of Illinois Press, 2001.

Lindgren, James M. "'Let Us Idealize Old Types of Manhood': The New Bedford Whaling Museum, 1903-1941." *The New England Quarterly* 72, no. 2 (1999): 163-206.

Lund, Judith Navas. *Whaling Masters and Whaling Voyages Sailing form American Ports, A Compilation of Sources*. New Bedford: Ten Pound Island Book Co., 2001.

Mawer, Granville Allen. *Ahab's Trade: The Saga of South Seas Whaling*. New York: St. Martin's Press, 1999.

McCullin, Thomas Austin. "Industrialization and Social Change in a Nineteenth Century Port City: New Bedford, Massachusetts, 1865-1900." Ph.D. Diss., University of Wisconsin, Madison, 1976.

———. "Lost Alternative: The Urban Industrial Utopia of William D. Howland." *The New England Quarterly* 55, no. 1 (1982): 25-38.

McCurry, Stephanie. *Masters of Small Worlds: Yeoman Households, Gender Relations, & the Political Culture of the Antebellum South Carolina Low Country*. New York: Oxford University Press, 1995.

Bibliography

McGuane, James. *The Hunted Whale*. New York: W. W. Norton, 2013.

McMullin, Thomas A. "Overseeing the Poor: Industrialization and Public Relief in New Bedford, 1865–1900." *Social Service Review* 65, no. 4 (1991): 548–63.

Miller, Marla R., and Laura A. Miller. *A Generous Sea: Native Hawaiians, Pacific Islanders, and the Jewish Community in New Bedford Whaling & Whaling Heritage*. Boston: Northeast Region Ethnography Program, National Park Service, 2016.

Motta, Arthur P., Jr. "Seal of the City of New Bedford, Massachusetts." City of New Bedford Office of Tourism & Marketing, 2013.

Mulderink, Earl F. III. "'A Burning and Shining Light': Prosperity and Enlightened Governance in Antebellum New Bedford." In *New Bedford's Civil War*. New York: Fordham University, 2012.

Nichols, Peter. *Oil and Ice: A Story of Arctic Disaster and the Rise and Fall of America's Last Whaling Dynasty*. New York: Penguin, 2009.

Orvell, Miles. *The Real Thing: Imitation and Authenticity in American Culture, 1880–1940*. Chapel Hill: University of North Carolina Press, 1989.

Paterson, Michael. *Inside Dickens' London*. Cincinnati: David and Charles Limited, 2011.

Philbrick, Nathaniel. *In the Heart of the Sea: The Tragedy of the Whaleship Essex*. New York: Viking, 2000.

Roberts, Arthur R. "A Quaker Understanding of Jesus Christ." *Quaker Religious Thought* 93, no. 4 (1999): 9–23.

Rodrigues, Rose Pearl. "Occupational Mobility of Portuguese Males in New Bedford, Massachusetts: 1870 to 1900." Ph.D. Diss., New School for Social Research, 1990.

Rosenbloom, Joshua L. "The Challenges of Economic Maturity: New England, 1880–1940." *NBER Papers on The Program on the Development of the American Economy*. The National Bureau of Economic Research, February 1999, https://www.nber.org/papers/h0113.pdf.

Sandler, Martin W. *Trapped In Ice!* New York: Scholastic Non-Fiction, 2006.

Schooler, Lynn. *The Last Shot*. New York: Harpers Collins, 2005.

Scott Institute for Energy Innovation. "What fuel did Pittsburgh's first street lamps consume?" Interview with Joel Tarr. *Energy Bite*. Podcast audio. April 4, 2017. https://energybite.org/page/2/.

Shoemaker, Nancy. *Living with Whales: Documents and Oral Histories of Native New England Whaling History*. Amherst: University of Massachusetts Press, 2014.

Silvia, Joe. "Historical Personages of New Bedford: James E. Reed." *Newbedfordguide.com*. March 9, 2013, https://www.newbedfordguide.com/james-e-reed/2013/03/09.

Sokalski, J.A. *Pictorial Illusionism: The Theatre of Steele MacKaye*. Montreal: McGill-Queen's University Press, 2007.

———. "The Theatre of Steele MacKaye: Pictorial Illusion on the American Stage." Ph.D. Diss, University of Toronto, 1997.

Songini, Marc. *The Lost Fleet: A Yankee Whaler's Struggle Against the Confederate Navy and Arctic Disaster*. New York: St. Martin's Press, 2007.

Starbuck, Alexander. *History of the American Whale Fishery*. Secaucus, NJ: Castle Books, 1989.

"Streetlamps." *Old House Journal*, July/August, 1988.

Thale, Christopher. "Calumet River System." *Encyclopedia of Chicago*. Chicago Historical Society. Accessed October 22, 2019. http://www.encyclopedia.chicagohistory.org/pages/203.html.

Vogel, Mike. "The Light from the Whale." *The Keeper's Log*, Fall 2014.

Warrin, Donald. *So Ends This Day: The Portuguese in American Whaling 1765–1927*. East Providence, RI: Signature Printing, 2010.

Williford, James. "Whaling the Old Way." *HUMANITIES: The Magazine of the National Endowment for the Humanities* 31, no. 2 (2010): accessed October 22, 2019. https://www.neh.gov/humanities/2010/marchapril/feature/whaling-the-old-way.

Wolfein, Seymour Louis. *The Decline of a Cotton Textile City*. New York: Columbia University Press, 1944.

Index

Numbers in **_bold italics_** indicate pages with illustrations

Almy, James T. 13–16, 20, 22
Ambergris 19, 35
Arctic Disaster of 1871 51, 56–66, 71–72; lithographs **_64_**–65; press coverage 65–66, 71; warnings about 57, 58–61
Arctic Whaling Exhibit Company 90, 92–94, 99, 116, 157–158, 165, 167; vs. *Tug James Hay* (admiralty court case) 137–140, 143, 157–158

Baleen (whalebone) 52–53; and corsets 53, **_54_**, 128
Bartlett, George F. 86–87, 90, 157–158, 167
Brennan, Kate 67–69, 72, 77–79

Charles W. Morgan 97, 98, 170
Chicago World's Fair of 1893 *see* World's Columbian Exposition
Columbian Exposition *see* World's Columbian Exposition
Crane, Fitch 98–104, 106, 119, 122, 133, 157, **_168_**, 169
Crapo, William W. 76, 79, 154–**_155_**, 167
cutting in 32, **_34_**, 85

Dowden, Capt. James 50, 53, 56, 59–66, 86, 101, 116–119, 124–127, 130, 132–133, 169

Field Museum 156, 158–162, 166
Florida (Confederate raider) 44, 47
Florida (whaling bark) 23–25, 27, 30–36, 55, 60–61
Franklin 85–87, 90, 104
Frazier, Capt. David 60–63, 66

Gifford, Capt. Daniel 1, 47–**_48_**, 96–98, 102–103, 105–106, 109–110, 112, 116, 118, 132–134, 137, 169

harpoon **_26_**, 27–31, 84, 90–91, 101, 106, 108, 120–121, 130–131, 146, 149, 160; Temple toggle iron 27
Howland, Ellis 154–156, 167
Howland, Matthew 41–42, 44, 47, 49–50, 51, 73, 75, 77, 79
Howland Mission Chapel 79

J. & W.R. Wing 74, 97–98, 157
Japan 53, 56, 58, 61

Kanaka, James "Jimmy" (Fiji King) 113, 130, 135, 147
Kempton, David 40–50, 85, 96–97

Lucem Diffundo 15–17, 19–20, 22, 135, 166

MacKaye, Steele 93–94, 115, 119, 122–123, 127, 135–136, 152–153, 169
Moby Dick (Melville) 17, 20, 24, 28, 33, 36, 40, 94

Nantucket sleighride 28–29, 31
National Museum *see* Smithsonian Institution
New Bedford: Board of Trade 76, 79, 80, 86–87, 90, 95, 98, 101, 118, 123, 157, 165; city seal 13–16, **_15_**, 19–20, 22; and cotton mill industry 67–81, 134–135, 154; and immigrants 67–69, 72–73, 77–80

Old Dartmouth Historical Society (ODHS) 49, 154–155, 167, 170

Peabody Museum (Peabody Essex Museum) 162
Peter the Whaler (Kingston) 30, 37, **_38_**, 39
Potomska Mill 68, 70–72, 154
Progress: abandonment after the fair 156–157, 162–164; and the Arctic Disaster 51, 59–60, 62–**_64_**, 66, 92, 101, 107, 125–126; arrival in Chicago 115–123, **_117_**; in Buffalo 105–106; departure from New Bedford 5, **_6_**, 7, 95, 98; in Detroit 106–109; discharge of crew 112–114; end in Chicago 7, **_8_**, 9, 155, 164, **_165_**; in Montreal 98–101, 105; as museum of "marine curiosities" 120–124, 128, 130–131, 134, 146, 149–150, 158; in Racine 109–114; sale of 86–87, 90, 100, 153, 157; sinking in Chicago River 135–140, 157; at State Street Bridge 124–130, 135; at World's Columbian Exposition 141- **_142_**, 145–153, **_147_**, **_151_**; writeups in fair guides 145–149

193

Index

Quakers (Religious Society of Friends) 20–22, 73, 78

Religious Society of Friends *see* Quakers

Shenandoah 45, **46**, 57
Smithsonian Institution (National Museum) 84, 91, 131–132, 160
Spectatorium **93**–94, 115, 122, 127, 135–136, 153, 169
sperm oil 17–18, **21**, 33, 37, 51; and lighthouses 18–20
sperm whales 17, 21–22, 23, 25, **26**, 28, **31**, 35–36, 52–53
spermaceti candles 17–18, 20, 22, 37, 52
Stone Fleet 42–44, **43**, 51

"There She Blows" or *The Story of the Progress* (Coffin) 143–144
trying out 33–**34**, 52, 119

Union League Club of Chicago 88–89, 90, 94, 150, 156, 158

walrus hunting 57–58
Wamsutta Mill 67–**70**, 72, 75, 77, 79, 154
Weaver, Henry 82–84, 87, 93–94, 98–100, 115, 119, 122–124, 131, 142–144, 165, 168; financial difficulties 150, 152–153, 157–158; interest in history 88–89
whale oil 17, 19, 22, 33, 37, 44, 51–52, 101
whalebone *see* baleen
Whalemen's Shipping List 23, 48, 65, 71, 73–74, 87, 133
whaling industry 10, 19, 24, 37, 40–41, 52; decline 42, 49, 50–52, 71–75, 81, 94; importance 10, 17, 19, 21–22, 73, 95; in popular culture 36–37, **38**, 39, 105, 108, 126–127; and previous expositions 84, 131; and whaling agents 41–42, 47, 49–50, 66, 85
Williams, Eliza 23–25, 28, 31–36, 55, 61, 63
Williams, William Fish 55, 59, 61, 63, 65, 107, 169
World's Columbian Exposition (Chicago World's Fair of 1893), (Columbian Exposition) 2, 11, 82–85, 88–89, 108, 115, 123, 131, 145–147, 152, 164